Intelligent Control Systems with LabVIEW™

T0185153

Pedro Ponce-Cruz • Fernando D. Ramírez-Figueroa

Intelligent Control Systems with LabVIEW™

 Springer

Pedro Ponce-Cruz, Dr.-Ing.
Fernando D. Ramírez-Figueroa, Research Assistant to Doctor Ponce

Instituto Tecnológico de Estudios Superiores de Monterrey
Campus Ciudad de México
Calle del Puente 222
Col. Ejidos de Huipulco Tlalpan
14380, México, D.F.
México

pedro.ponce@itesm.mx

ISBN 978-1-4471-6115-8   ISBN 978-1-84882-684-7 (eBook)
DOI 10.1007/978-1-84882-684-7
Springer London Dordrecht Heidelberg New York

British Library Cataloguing in Publication Data
A catalogue record for this book is available from the British Library

Additional material to this book can be downloaded from http://extras.springer.com.

This book is dedicated to my mother and son with love.

Pedro Ponce

To my parents Virginia and Fernando because this would not be possible without their unconditional love and support, and to my advisor and friend, Pedro for giving me the opportunity to prove myself, and to all those who have accompanied me along this wonderful journey of knowledge and creation.

David Ramirez

Preface

Control systems are becoming more important every day. At the beginning, the industry used sequential controls for solving a lot of industrial applications in control systems, and then the linear systems gave us a huge increase in applying automatic linear control on industrial application. One of the most recent methods for controlling industrial applications is intelligent control, which is based on human behavior or concerning natural process.

Nowadays, the topic of intelligent control systems has become more than a research subject to the industry. The number of industrial applications is growing every day, faster and faster. Thus, new software and hardware platforms are required in order to design and develop intelligent control systems. The challenge for these types of systems is to have a novel platform, which allows designing, testing and implementing an intelligent controller system in a short period of time. For the industry and academy, LabVIEW™ is one of the most important software platforms for developing engineering applications and could be connected with different hardware systems, as well as running standalone programs for simulating the controller's performance (validating the controller by simulation then implementing it). In addition, LabVIEW is a graphical program that is very easy to learn.

Taking into account these advantages, the software platform described in this book is LabVIEW from National Instruments™. The book is divided into 7 chapters and gives all the information required for designing and implementing an intelligent controller.

Chapter 1 provides an introduction to basic intelligent control concepts and concludes by applying LabVIEW for implementing control systems. Chapter 2 covers in deep detail the fuzzy logic theory and implementation. This chapter starts with fundamental fuzzy logic theory for supporting the most important fuzzy logic controllers implemented using LabVIEW.

Chapter 3 deals with artificial neural networks. In this chapter a complete set of tools for implementing artificial neural networks is presented. Basic examples of neural networks, such as perceptron, allow the students to understand the most important topologies in artificial neural networks for modeling and controlling systems. In Chap. 4 the reader can find neuro-fuzzy controllers, which combine the fuzzy inference systems with an artificial neural network topology. Thus, the neuro-fuzzy controllers are an interesting option for modeling and controlling industrial applications. Chapter 5 discusses genetic algorithms, which are representations of the natural selection process. This chapter also examines how generic algorithms can be used as optimization methods. Genetic programming is also explained in detail.

Chapters 6 and 7 show different algorithms for optimizing and predicting that could be combined with the conventional intelligent system methodologies presented in the previous chapters such as fuzzy logic, artificial neural networks and neuro-fuzzy systems. The methods presented in Chaps. 6 and 7 are: simulated annealing, fuzzy clustering means, partition coefficients, tabu search and predictors.

Supplemental materials supporting the book are available in the companion DVD. The DVD includes all the LabVIEW programs (VIs) presented inside the book for intelligent control systems.

This book would never have been possible without the help of remarkable people who believed in this project. I am not able to acknowledge all of them here, but I would like to thank Eloisa Acha, Gustavo Valdes, Jeannie Falcon, Javier Gutierrez and others at National Instruments for helping us to develop a better book.

Finally, I would like to thank the Instituto Tecnológico de Monterrey campus Ciudad de México for supporting this research project. I wish to remember all my friends and colleagues who gave me support during this research journey.

ITESM-CCM   *Dr. Pedro Ponce-Cruz*
México City

Contents

Chapter 1
Intelligent Control for LabVIEW

1.1 Introduction

Intelligent control techniques that emulate characteristics of biological systems offer opportunities for creating control products with new capabilities. In today's competitive economic environment, these control techniques can provide products with the all-important competitive edge that companies seek. However, while numerous applications of intelligent control (IC) have been described in the literature, few advance past the simulation stage to become laboratory prototypes, and only a handful make their way into products. The ability of research to impact products hinges not so much on finding the best solution to a problem, but on finding the right problem and then solving it in a marketable way [1].

The study of intelligent control systems requires both defining some important expressions that clarify these systems, and also understanding the desired application goals. The following definitions show the considerable challenges facing the development of intelligent control systems.

Intelligence is a mental quality that consists of the abilities to learn from experience, adapt to new situations, understand and handle abstract concepts, and use knowledge to manipulate one's environment [2]. We can define *artificial intelligence* as the ability of a digital computer or computer-controlled robot to perform tasks commonly associated with intelligent beings [2].

Thus, IC is designed to seek control methods that provide a level of intelligence and autonomy in the control decision that allows for improving the system performance. As a consequence, IC has been one of the fastest growing areas in the field of control systems over the last 10 years. Even though IC is a relatively new technique, a huge number of industrial applications have been developed. IC has different tools for emulating the biological behavior that could solve problems as human beings do. The main tools for IC are presented below:

- *Fuzzy logic systems* are based on the experience of a human operator, expressed in a linguistic form (normally *IF–THEN* rules).

P. Ponce-Cruz, F. D. Ramirez-Figueroa, *Intelligent Control Systems with LabVIEW™*
© Springer 2010

- *Artificial neural networks* emulate the learning process of biologic neural networks, so that the network can learn different patterns using a training method, supervised or unsupervised.
- *Evolutionary methods* are based on evolutionary processes such as natural evolution. These are essentially optimization procedures.
- *Predictive methods* are mathematical methods that provide information about the future system behavior.

Each one has advantages and disadvantages, but some of the disadvantages can be decreased by combining two or more methods to produce one system (hybrid systems). As an example, in the case of fuzzy logic, we can combine this method with neural networks to obtain a neuro-fuzzy system. For instance, the adaptive neural-based fuzzy inference system (ANFIS) was proposed in order to utilize the best part of fuzzy logic inference using an adaptive neural network topology [3].

Different authors have presented many hybrid systems, but the most important and useful combinations are [4]:

- Neural networks combined with genetic algorithms [5].
- Fuzzy systems combined with genetic algorithms [6].
- Fuzzy systems combined with neural networks [7].
- Various other combinations have been implemented [8, 9].

Since fuzzy logic was first presented by Prof. Lotfi A. Zadeh, the number of fuzzy logic control applications has increased dramatically. For example, in a conventional proportional, integral, and differential (PID) controller, what is modeled is the system or process being controlled, whereas in a fuzzy logic controller (FLC), the focus is the human operator's behavior. In the PID, the system is modeled analytically by a set of differential equations, and their solution tells the PID controller how to adjust the system's control parameters for each type of behavior required. In the fuzzy controller, these adjustments are handled by a fuzzy rule-based expert system, a logical model of the thinking processes a person might go through in the course of manipulating the system. This shift in focus from the process to the person involved changing the entire approach to automatic control problems [10].

The search has been ongoing for a controller, of a black box type, which can be simply plugged into a plant, where control is desired; thus, the controller takes over from there and sorts everything else out [10].

IC is a good solution for processes where the mathematical model that describes the system is known only partially. In fact, the PID controller is one of the most functional solutions used nowadays, because it requires a very short time for implementation and the tuning techniques are well known. We show in this book how fuzzy systems can be used to tune direct and adaptive fuzzy controllers, as well as, how these systems can be used in supervisory control.

Although the IC is more complex in structure than the PID controller, the IC gives a better response if the system changes to a different operation point. It is well known that linear systems are designed for working around the operation point. In the case of IC, we will be able to design controllers that work outside the opera-

Fig. 1.1 Basic sets for
obtaining IC systems

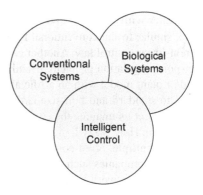

tion point. A global position in control theory of IC is shown in Fig. 1.1, in which different sets intersect in the IC area.

As it is presented, IC systems are in contrast to analytical control, because soft computing methodologies mimic consciousness and cognition in several important ways:

- To learn from experience.
- To be able to universalize into domains where direct experience is absent.
- To run into parallel computer architectures, which simulate biological processes.
- To perform mapping from inputs to the outputs faster than inherently serial analytical representations.

The trade off, however, is a decrease in accuracy. If a tendency towards imprecision can be tolerated, then it should be possible to expand the range of the applications even to those problems where the analytical and mathematical representations are readily available [11].

1.2 Intelligent Control in Industrial Applications

The number of industrial applications that use IC systems is rapidly increasing, where one can find IC systems in both large and small industrial applications. Another growing area of IC applications is developing household appliances, which are small but complex control systems. Many systems that use fuzzy logic or neural networks for control apply these techniques to solve problems that fall outside the domain of conventional feedback control, e. g., in the case of a washer machine it is easier to control the duty cycle by a FLC than a PID controller. When we view fuzzy or neural control as only a non-linear counterpart of conventional feedback control techniques, the possibilities of using IC are reduced. Thus, a narrow conceptual view of IC system application leads to designers not appreciating or recognizing new areas of opportunities. If you use only the IC systems as a conventional controller the difference is quite small. For instance, using a FLC as a PID controller with the error and the change in error as inputs, the fuzzy controllers look similar to the conventional PID controller except that fuzzy control provides a non-linear control law. Another case is the use of a neural network applied to the set-point regulation problem, usually by replacing a conventional controller's law and/or plant model with an artificial neural network. However, if we apply IC systems to standard and non-standard techniques we could handle high-level control systems. Let us imagine the control system of the train developed in Sendai, Japan by Hitachi. Here fuzzy logic was used to select the notch position that will best satisfy the multiple, often-conflicting objectives. An additional example is that many Japanese companies such as Matsushita, Sanyo, Hitachi, and Sharp, have incorporated neural network technology into a product known as the kerosene fan heater. In Sanyo's heater, a neural network learns the daily usage pattern of the consumer, thus allowing the heater to automatically start to preheat in advance [12]. For many industrial applications one could complement the conventional controllers by an intelligent controller generating a new one, rather than using IC alone. The industrial challenge is focused on developing control systems that are capable of adapting to rapidly changing environments and on improving their performance based on their experience. In other words, modern control systems are being developed that are capable of learning to improve their performance over time (to learn) much like humans do [4].

1.3 LabVIEW

LabVIEW is a graphical control, test, and measurement environment development package. Laboratory Virtual Instrument Engineering Workbench (LabVIEW) is a graphical programming language that uses icons instead of lines of text to create programs. It is one of the most widely used software packages for test, measurement, and control in a variety of industries. LabVIEW was first launched in 1986, but has continued to evolve and extend into targets seemingly out of reach before for such packages, including field programmable gate arrays (FPGAs), sensors, microcontrollers and other embedded devices. The introduction of the express virtual instruments (VIs) allow designers faster and easier development of block diagrams for any type of data acquisition, analysis, and control application. As a result, Lab-VIEW simplifies the scientific computation, process control, research, industrial application and measurement applications, since it has the flexibility of a programming language combined with built-in tools designed specifically for test, measurement, and control. By using the integrated LabVIEW environment, it is now easier to interface with real-world signals, analyze data for meaningful information, and share results [13, 14].

As we know, there are numerous programs on the market for controlling, analyzing, and processing signals but LabVIEW has a big advantage in its graphical user interface (GUI). LabVIEW was preferred over other programs by a wide margin for its easy-to-use GUI capabilities. This feature is an integral part of the LabVIEW

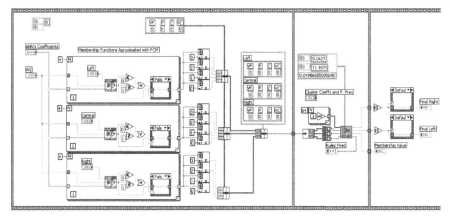

Fig. 1.2 Block diagram from Intelligent Control Toolkit

VI structure for code development. The other two criteria for which LabVIEW was slightly preferred included its easier-to-use real-time integration tools and help resources [14]. These advantages were taken into account for selecting LabVIEW as the main platform in the Intelligent Control Toolkit design. Thus, all of the material presented in this book was generated by VIs, which are the basic programs in Lab-VIEW [13]. "Ever since LabVIEW shipped, it had more recognition than NI," said Kodosky who is one of the founders of National Instruments [13].

The programs that simulate virtual instruments created in LabVIEW are called VIs. The VI has three basic components: the front panel, the block diagram, and the icon connector. The control panel is the user interface and the code is inside the block diagram that contains the graphical code. Also, one could include a subVI that must have an icon and a connector pane, where the subVI is generated as a VI. Figures 1.2 and 1.3 show the block diagram and the front panel from a VI example of the Intelligent Control Toolkit [15].

In the front panel, one can add the number of inputs and outputs that the system requires. The basic elements inside the front panel can be classified by controls and indicators. The general type of numerical data can be integers, floating, and complex numbers. Another type of data is the Boolean, useful in conditional systems (true or false), as well as strings, which are a sequence of ASCII characters that give a platform-independent format for information and data [16].

Using control loops, it is possible to repeat a sequence of programs or to enter the program conditions. The control loops used are shown in Fig. 1.4. It is also possible to analyze the outputs of the intelligent systems by a waveform chart or other block, by plotting the output data. Figure 1.5 shows a waveform chart used for analyzing output signals.

In the case of inputs, which are representations of physical phenomenon, one could obtain the information by a data acquisition system. One of the main goals in the data acquisition system for obtaining a successful system is the selection of the system, as well as, the transducer and sensors. The data acquisition system plays

Fig. 1.3 Front panel. Adapted from [15]

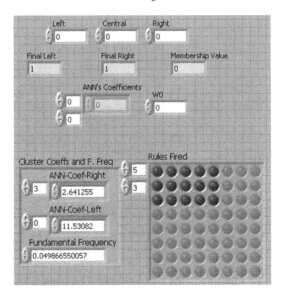

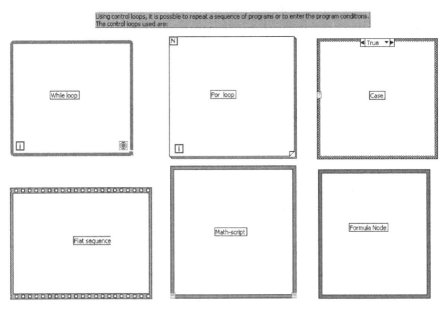

Fig. 1.4 Control block loops

a key role in the control system design. Nowadays National Instruments is one of the most important companies in the world for providing excellent acquisition systems. Different acquisition systems are shown in Fig. 1.6.

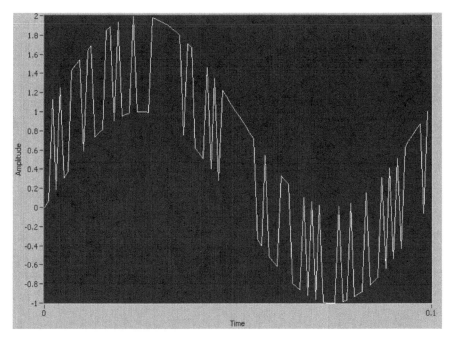

Fig. 1.5 Waveform chart

Fig. 1.6a,b Acquisition systems developed by NI [16]. **a** CompactRIO™. **b** NI USB DAQ

References

1. Chiu S (1997) Developing commercial applications of intelligent control. IEEE Control Syst Mag 17(2):94–100
2. Encyclopedia Britannica (2009) Intelligence. http://www.britannica.com/EBchecked/topic/289766/human-intelligence. Accessed on 22 March 2009
3. Jang J-SR (1993) ANFIS: Adaptive network-based fuzzy inference systems. IEEE Trans Syst Man Cybernet 23:665–685

4. Karr CL (2003) Control of a phosphate processing plant via a synergistic architecture. Eng Appl Artif Intell 6(1):21–30
5. Van Rooij AJF (1996) Neural network training using genetic algorithms. World Scientific, River Edge, NJ
6. Sanchez E, Shibata T, Zadeh LA (1997) Genetic algorithms and fuzzy logic systems. Advances in fuzzy systems: application and theory, Vol. 7. World Scientific, River Edge, NJ
7. Kosko B (1991) Neural networks and fuzzy systems: a dynamical systems approach to machine intelligence. Prentice Hall, New York
8. Goonatilake S, Khebbal S (eds) (1996) Intelligent hybrid systems. Wiley, New York
9. Medsker LR (1995) Hybrid intelligent systems. Kluwer, Dordrecht
10. Schwartz DG, Klir GJ (1992) Fuzzy logic flowers in Japan. IEEE Spectrum 29(7):32–35
11. Zilouchian A, Jamshidi M (2001) Intelligent control systems using soft computing methodologies. CRC, Boca Raton, FL
12. Warwick K (1998) Recent developments in intelligent control. IEE Colloquium on Updates on Developments in Intelligent Control, Oct 1998, pp 1/1–1/4
13. Josifovska S (2003) The father of LabVIEW. IEE Rev 49(9):30–33
14. Kehtarnavaz N, Gope C (2006) DSP system design using LabVIEW and Simulink: a comparative evaluation. Proceedings of IEEE International Conference on Acoustics, Speech and Signal Processing (ICASSP), Toulouse, France, 14–19 May 2006, Vol. 2, pp II–II
15. ITESM Mexico (2007) Intelligent Control Toolkit for LabVIEW. US Patent Application 61/197,484
16. National Instruments (2009) http://www.ni.com. Accessed on 22 March 2009

Chapter 2
Fuzzy Logic

2.1 Introduction

The real world is complex; this complexity generally arises from uncertainty. Humans have unconsciously been able to address complex, ambiguous, and uncertain problems thanks to the gift of thinking. This thought process is possible because humans do not need the complete description of the problem since they have the capacity to reason approximately. With the advent of computers and their increase in computation power, engineers and scientists are more and more interested in the creation of methods and techniques that will allow computers to reason with uncertainty.

Classical set theory is based on the fundamental concept of a *set*, in which individuals are either a member or not a member. A sharp, crisp, and ambiguous distinction exists between a member and a non-member for any well-defined set of entities in this theory, and there is a very precise and clear boundary to indicate if an entity belongs to a set. Thus, in classical set theory an element is not allowed to be in a set (1) or not in a set (0) at the same time. This means that many real-world problems cannot be handled by classical set theory. On the contrary, the fuzzy set theory accepts partial membership values $\mu_f \in [0, +1]$, and therefore, in a sense generalizes the classical set theory to some extent.

As Prof. Lotfi A. Zadeh suggests by his *principle of incompatibility*: "The closer one looks at a real-world problem, the fuzzier becomes the solution," and thus, imprecision and complexity are correlated [1]. Complexity is inversely related to the understanding we can have of a problem or system. When little complexity is presented, closed-loop forms are enough to describe the systems. More complex systems need methods such as neural networks that can reduce some uncertainty. When systems are complex enough that only few numerical data exist and the majority of this information is vague, fuzzy reasoning can be used for manipulating this information.

2.2 Industrial Applications

The imprecision in fuzzy models is generally quite high. However, when precision is apparent, fuzzy systems are less efficient than more precise algorithms in providing us with the best understanding of the system. In the following examples, we explain how many industries have taken advantage of the fuzzy theory [2].

Example 2.1. Mitsubishi manufactures a fuzzy air conditioner. While conventional air conditioners use on/off controllers that work and stop working based on a range of temperatures, the Mitsubishi machine takes advantage of fuzzy rules; the machine operates smoother as a result. The machine becomes mistreated by sudden changes of state, more consistent room temperatures are achieved, and less energy is consumed. These were first released in 1989.   □

Example 2.2. Fisher, Sanyo, Panasonic, and Canon make fuzzy video cameras. These have a digital image stabilizer to remove hand jitter, and the video camera can determine the best focus and lightning. Fuzzy decision making is used to control these actions. The present image is compared with the previous frame in memory, stationary objects are detected, and its shift coordinates are computed. This shift is subtracted from the image to compensate for the hand jitter.   □

Example 2.3. Fujitec and Toshiba have a fuzzy scheme that evaluates the passenger traffic and the elevator variables to determine car announcement and stopping time. This helps reduce the waiting time and improves the efficiency and reliability of the systems. The patent for this type of system was issued in 1998.   □

Example 2.4. The automotive industry has also taken advantage of the theory. Nissan has had an anti-lock braking system since 1997 that senses wheel speed, road conditions, and driving pattern, and the fuzzy ABS determines the braking action, with skid control [3].   □

Example 2.5. Since 1988 Hitachi has turned over the control of the Sendai subway system to a fuzzy system. It has reduced the judgment on errors in acceleration and braking by 70%. The Ministry of International Trade and Industry estimates that in 1992 Japan produced about $2 billion worth of fuzzy products. US and European companies still lag far behind. The market of products is enormous, ranging from fuzzy toasters to fuzzy golf diagnostic systems.   □

2.3 Background

Prof. Lotfi A. Zadeh introduced the seminal paper on fuzzy sets in 1965 [4]. Since then, many developments have taken place in different parts of the world. Since the 1970s Japanese researchers have been the primary force in the implementation of fuzzy theory and now have thousands of patents in the area.

The world response to fuzzy logic has been varied. On the one hand, western cultures are mired with the *yes* or *no*, *guilty* or *not guilty*, of the binary Aristotelian logic world and their interpretation of the fuzziness causes a conflict because they are given a negative connotation. On the other hand, Eastern cultures easily accommodate the concept of fuzziness because it does not imply disorganization and imprecision in their languages as it does in English.

2.3.1 Uncertainty in Information

The uncertainties in a problem should be carefully studied by engineers prior to selecting an appropriate method to represent the uncertainty and to solve the problem. Fuzzy sets provide a way that is very similar to the human reasoning system. In universities most of the material taught in engineering classes is based on the presumption that knowledge is deterministic. Then when students graduate and enter "the real world," they fear that they will forget the correct formula.

However, one must realize that all information contains a certain degree of uncertainty. Uncertainty can arise from many factors, such as complexity, randomness, ignorance, or imprecision. We all use vague information and imprecision to solve problems. Hence, our computational methods should be able to represent and manipulate fuzzy and statistical uncertainties.

2.3.2 Concept of Fuzziness

In our everyday language we use a great deal of vagueness and imprecision, that can also be called fuzziness. We are concerned with how we can represent and manipulate inferences with this kind of information. Some examples are: a person's size is *tall*, and their age is classified as *young*.

Terms such as *tall* and *young* are fuzzy because they cannot be crisply defined, although as humans we use this information to make decisions. When we want to classify a person as tall or young it is impossible to decide if the person is in a set or not. By giving a *degree* of pertinence to the subset, no information is lost when the classification is made.

2.4 Foundations of Fuzzy Set Theory

Mathematical foundations of fuzzy logic rest in fuzzy set theory, which can be seen as a generalization of classical set theory. Fuzziness is a language concept; its main strength is its vagueness using symbols and defining them.

Consider a set of tables in a lobby. In classical set theory we would ask: Is it a table? And we would have only two answers, *yes* or *no*. If we code *yes* with a 1 and *no* with a 0 then we would have the pair of answers as {0,1}. At the end we would collect all the elements with 1 and have the set of tables in the lobby.

We may then ask what objects in the lobby can *function* as a table? We could answer that tables, boxes, desks, among others can function as a table. The set is not uniquely defined, and it all depends on what we mean by the word *function*. Words like this have many shades of meaning and depend on the circumstances of the situation. Thus, we may say that the set of objects in the lobby that can function as a table is a *fuzzy set*, because we have not crisply defined the criteria to define the *membership* of an element to the set. Objects such as tables, desks, boxes may function as a table with a certain degree, although the fuzziness is a feature of their representation in symbols and is normally a property of models, or languages.

2.4.1 Fuzzy Sets

In 1965 Prof. Lotfi A. Zadeh introduced *fuzzy sets*, where many degrees of membership are allowed, and indicated with a number between 0 and 1. The point of departure for fuzzy sets is simply the generalization of the valuation set from the pair of numbers $\{0,1\}$ to all the numbers in $[0,1]$. This is called a *membership function* and is denoted as $\mu_A(x)$, and in this way we have fuzzy sets.

Membership functions are mathematical tools for indicating flexible membership to a set, modeling and quantifying the meaning of symbols. They can represent a subjective notion of a vague class, such as chairs in a room, size of people, and performance among others. Commonly there are two ways to denote a fuzzy set. If X is the universe of discourse, and x is a particular element of X, then a fuzzy set A defined on X may be written as a collection of ordered pairs:

$$A = \{(x, \mu_A(x))\} \qquad x \in X, \tag{2.1}$$

where each pair $(x, \mu_A(x))$ is a *singleton*. In a crisp set singletons are only x, but in fuzzy sets it is two things: x and $\mu_A(x)$. For example, the set A may be the collection of the following integers, as in (2.2):

$$A = \{(1, 1.0), (3, 0.7), (5, 0.3)\} . \tag{2.2}$$

Thus, the second element of A expresses that 3 belongs to A to a degree of 0.7. The *support set* of a fuzzy set A is the set of elements that have a membership function different from zero. Alternative notations for the fuzzy sets are *summations* or *integrals* to indicate the *union* of the fuzzy set, depending if the universe of discourse is *discrete* or *continuous*. The notation of a fuzzy set with a discrete universe of discourse is $A = \sum_{x_i \in X} \mu_A(x_i)/x_i$ which is the union of all the singletons. For a continuous universe of discourse we write the set as $A = \int_X \mu_A(x)/x$, where the integral sign indicates the union of all $\mu_A(x)/x$ singletons.

Now we will show how to create a triangular membership function using the Intelligent Control Toolkit for LabVIEW (ICTL). This triangular function must be between 0 and 3 with the maximum point at 1.5; we can do this using the **triang-function.vi**. To evaluate and graph the function we must use a 1D array that can be easily created using the **rampVector.vi**. We can find this VI (as shown in Fig. 2.1)

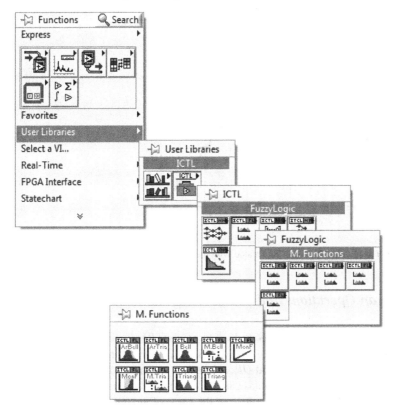

Fig. 2.1 Fuzzy function location on the ICTL

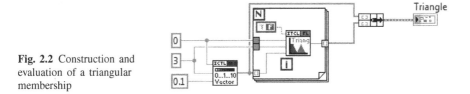

Fig. 2.2 Construction and evaluation of a triangular membership

in the fuzzy logic palette of the toolkit. The block diagram of the program that will create and evaluate the triangular function is shown in Fig. 2.2. The triangular function will be as the one shown in Fig. 2.3.

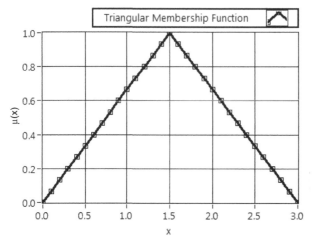

Fig. 2.3 Triangular membership function created with the ICTL

2.4.2 *Boolean Operations and Terms*

The two-valued logic is called *Boolean algebra*, named after George Boole, a nine-teenth century mathematician and logician. In this algebra there are only three basic logic operations: *NOT* ¬, *AND* ∧ and *OR* ∨. It is also common to use the symbols: −, ·, and +. Boolean algebraic formulas can be described by a truth table, where all the variables in the formula are the inputs and the value of the formula is the output. Conversely, a formula can be written from a truth table. For example the truth table for *AND* is shown in Table 2.1.

Complex Boolean formulas can be reduced to simpler equivalent ones using some properties. It is important to note that some rules of the Boolean algebra are the same as those of the ordinary algebra (e. g., $a \cdot 0 = 0$, $a \cdot 1 = a$), but others are quite different ($a + 1 = 1$). Table 2.2 shows the most important properties of Boolean algebra.

Table 2.1 Truth table of the *AND* Boolean operation

x	y	$x \wedge y$
0	0	0
0	1	0
1	0	0
1	1	1

Table 2.2 The most important properties of Boolean algebra

Laws	Formulas
Characteristics	$a \cdot 0 = 0, a \cdot 1 = a, a + 0 = a$ and $a + 1 = 1$
Commutative law	$a + b = b + a$ and $a \cdot b = b \cdot a$
Associative law	$a + b + c = a + (b + c) = (a + b) + c$
	$a \cdot b \cdot c = a \cdot (b \cdot c) = (a \cdot b) \cdot c$
Distributive law	$a \cdot (b + c) = a \cdot b + a \cdot c$
Idempotence	$a \cdot a = a$ and $a + a = a$
Negation	$\overline{\overline{a}} = a$
Inclusion	$a \cdot \overline{a} = 0$ and $a + \overline{a} = 1$
Absorptive law	$a + a \cdot b = a$ and $a \cdot (a + b) = a$
Reflective law	$a + \overline{a} \cdot b = a + b, a \cdot (\overline{a} + b) = a \cdot b$, and $a \cdot b + \overline{a} \cdot b \cdot c = a \cdot b + b \cdot c$
Consistency	$a \cdot b + a \cdot \overline{b} = a$ and $(a + b) \cdot \left(a + \overline{b}\right) = a$
De Morgan's law	$\overline{a \cdot b} = \overline{a} + \overline{b}$ and $\overline{a + b} = \overline{a} \cdot \overline{b}$

2.4.3 Fuzzy Operations and Terms

Operations such as *intersection* and *union* are defined through the *min* ($\wedge$) and *max* ($\vee$) operators, which are analogous to *product* and *sum* in algebra. Formally the min and max of an element, where $\equiv$ stands for "by definition," are denoted by (2.3) and (2.4):

$$\mu_a \wedge \mu_b = \min(\mu_a, \mu_b) \equiv \begin{cases} \mu_a \text{ if and only if } \mu_a \leq \mu_b \\ \mu_b \text{ if and only if } \mu_a > \mu_b \end{cases} \quad (2.3)$$

$$\mu_a \vee \mu_b = \max(\mu_a, \mu_b) \equiv \begin{cases} \mu_a \text{ if and only if } \mu_a \geq \mu_b \\ \mu_b \text{ if and only if } \mu_a < \mu_b \end{cases}. \quad (2.4)$$

The most important fuzzy operations are shown in Table 2.3. The following functions in (2.5) are two fuzzy sets, a triangular and a bell-shaped membership function:

$$\mu_{\text{triangle}}(x) = \begin{cases} \frac{2(x-1)}{7}; & 1 \leq x \leq \frac{9}{2} \\ -\frac{2(x-8)}{7}; & \frac{9}{2} \leq x \leq 8 \end{cases} \qquad \mu_{\text{bell}}(x) = \frac{1}{1 + \left|\frac{x-0.1}{3}\right|^6}. \quad (2.5)$$

The diagrams for the membership functions can be found in Fig. 2.4, a union between the triangular and bell functions is shown in Fig. 2.5, and an intersection is shown in Fig. 2.6. The bell function and the complement are shown in Fig. 2.7.

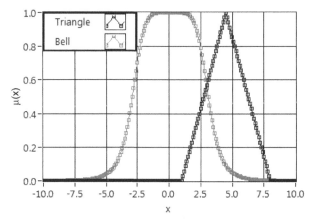

Fig. 2.4 Diagram of triangular and bell membership functions

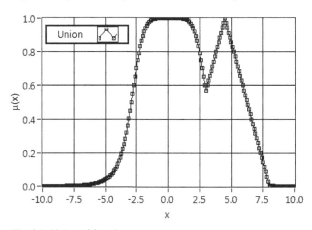

Fig. 2.5 Union of functions

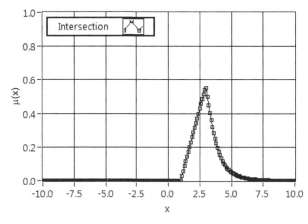

Fig. 2.6 Intersection of sets

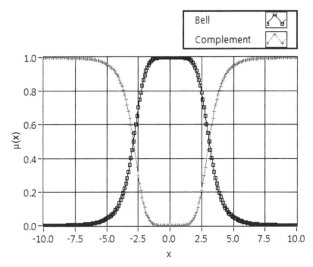

Fig. 2.7 Bell and complement of the bell function

Table 2.3 The most important fuzzy operations

Empty fuzzy set	It is empty if its membership function is zero everywhere in the universe of discourse.	$A \equiv \emptyset$ if $\mu_A(x) = 0, \forall x \in X$
Normal fuzzy set	It is normal if there is at least one element in the universe of discourse where its membership function equals one.	$\mu_A(x_a) = 1$
Union of two fuzzy sets	The union of two fuzzy sets A and B over the same universe of discourse X is a fuzzy set $A \cup B$ in X with a membership function which is the maximum of the grades of membership of every x and A and B. This operation is related to the *OR* operation in fuzzy logic.	$\mu_{A \cup B}(x) \equiv \mu_A(x) \vee \mu_B(x)$
Intersection of fuzzy sets	It is the minimum of the grades of every x in X to the sets A and B. The *intersection* of two fuzzy sets is related to the *AND*.	$\mu_{A \cap B}(x) \equiv \mu_A(x) \wedge \mu_B(x)$
Complement of a fuzzy set	The complement of a fuzzy set A is denoted as $\bar{A}$.	$\mu_{\bar{A}}(x) \equiv 1 - \mu_A(x)$
Product of two fuzzy sets	$A \cdot B$ denotes the product of two fuzzy sets with a membership function that equals the algebraic product of the membership function A and B.	$\mu_{A \cdot B}(x) \equiv \mu_A(x) \cdot \mu_B(x)$

Table 2.3 (continued)

Power of a fuzzy set	The β power of A (A^β) has the equivalence to linguistically modify the set with *VERY.*	$\mu_{A^\beta}(x) \equiv [\mu_A(x)]^\beta$
Concentration	Squaring the set is called *concentration CON.*	$\mu_{CON(\beta)}(x) \equiv (\mu_A(x))^2$
Dilation	Taking the square root is called *dilation* or *DIL.*	$\mu_{DIL(A)}(x) \equiv \sqrt{\mu_A(x)}$

2.4.4 Properties of Fuzzy Sets

Fuzzy sets are useful in performing operations using membership functions. Properties listed in Table 2.4 are valid for crisp and fuzzy sets, although some are specific for fuzzy sets only. Sets A, B, and C must be considered as defined over a common universe of discourse X.

All of these properties can be expressed using the membership function of the sets involved and the definitions of union, intersection and complement. De Morgan's law says that the intersection of the complement of two fuzzy sets equal the complement of their union. There are also some properties not valid for fuzzy sets such as the law of contradiction and the law of the excluded middle.

Table 2.4 The most important fuzzy properties

Double negation law	$\overline{(\bar{A})} = A$
Idempotency	$A \cup A = A \quad A \cap A = A$
Commutativity	$A \cap B = B \cap A \quad A \cup B = B \cup A$
Associative property	$(A \cup B) \cup C = A \cup (B \cup C)$ $(A \cap B) \cap C = A \cap (B \cap C)$
Distributive property	$A \cup (B \cap C) = (A \cup B) \cap (A \cup C) \, A \cap$ $(B \cup C) = (A \cap B) \cup (A \cap C)$
Absorption	$A \cap (A \cup B) = A$ $A \cup (A \cap B) = A$
De Morgan's laws	$\overline{A \cup B} = \bar{A} \cap \bar{B}$ $\overline{A \cap B} = \bar{A} \cup \bar{B}$

2.4.5 Fuzzification

This process is mainly used to transform a crisp set to a fuzzy set, although it can also be used to increase the fuzziness of a fuzzy set. A *fuzzifier function F* is used to control the fuzziness of the set. As an example the fuzzy set A can be defined with

Fig. 2.8 Diagram of the **Bell-Function.vi**

Bell-Function.vi

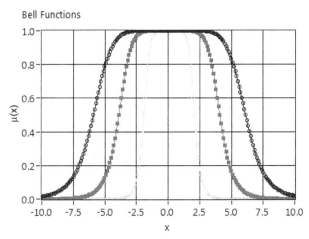

Fig. 2.9 Different forms of bell membership functions

the function in (2.6):

$$\mu_A\left(x\right) = \frac{1}{1+\left|\frac{x-c}{a}\right|^{2b}}, \qquad (2.6)$$

where x is any positive real number, and the parameters a, b, c shape the form of the bell membership function. The fuzzy set A can be written as:

$$A \equiv \int_x \left[\frac{1}{1+\left|\frac{x-c}{a}\right|^{2b}}\right]/x. \qquad (2.7)$$

This is an example of the bell function with different parameters using the ICTL. We can use the **Bell-Function.vi** (shown in Fig. 2.8) to create the membership functions. The a, b and c parameters can be changed and the form of the function will be different. The membership functions are shown in Fig. 2.9. The code that generates the membership functions is shown in Fig. 2.10. Basically, a 1D array is used to evaluate each one of the bell functions and generate their different forms.

Example 2.6. The productivity of people can be modeled using a bell function. It will increase depending on their age, then it will remain on the top for several years and it will decrease when the person reaches a certain age. This model is shown in the membership function (Triangular function with saturation) given in Fig. 2.11.   □

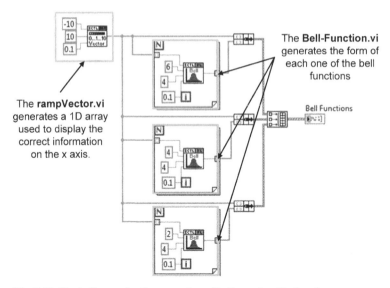

The **rampVector.vi**
generates a 1D array
used to display the
correct information
on the x axis.

The **Bell-Function.vi**
generates the form of
each one of the bell
functions

Bell Functions

Fig. 2.10 Block diagram for the generation of bell membership functions

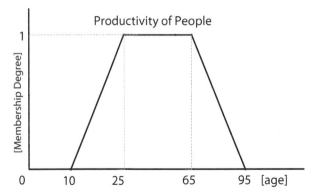

Fig. 2.11 Productivity of people fuzzy model

Why do not we select a conventional triangular membership function? The answer
is because triangular functions reach their maximum at only one number and we are
trying to model a range in which the productivity reaches its maximum. Thus, if we
use triangular functions we would be representing the maximum of the productivity
for a certain age of people (Fig. 2.12).

We can use a shoulder function to model a process, where after a certain level, the
degree of membership remains the same (Fig. 2.13). We may want to model the level

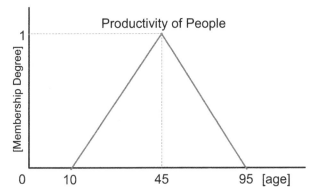

Fig. 2.12 Productivity of people modeled with a conventional triangular membership function

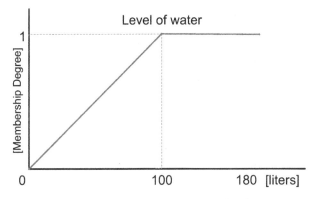

Fig. 2.13 Productivity modeled with a shoulder function

of water in a tank, which gets full after a certain number of liters are poured into the tank. Once we pass beyond that level, the degree of the level of water remains the same; the same happens if the tank is completely drained.

2.4.6 Extension Principle

This is a mathematical tool used to extend crisp mathematical notions and operations to the fuzzy realm, by fuzzifying the parameters of a function, resulting in computable fuzzy sets. Suppose that we have a function f that maps elements x_1, x_2, ..., x_n of a universe of discourse X to another universe of discourse Y, and a fuzzy set A defined on the inputs of f in (2.8):

$$y_1 = f(x_1)$$
$$y_2 = f(x_2)$$
$$\cdots \qquad A = \mu_A(x_1)/x_1 + \mu_A(x_2)/x_2 + \cdots + \mu_A(x_n)/x_n. \qquad (2.8)$$
$$y_n = f(x_n)$$

What could happen if the input of function f becomes fuzzy? Would the output be fuzzy? The extension principle then tells us that there is a fuzzy output given by (2.9):

$$B = f(B) = \mu_A(x_1)/f(x_1) + \mu_A(x_2)/f(x_2) + \cdots + \mu_A(x_n)/f(x_n), \quad (2.9)$$

where every single image of x_i under f becomes fuzzy to a degree $\mu_A(x_i)$. Most of the functions out there are *many-to-one*, meaning several x map the same y. We then have to decide which of the two or more membership values we should take as the membership value of the output. The extension principle says that the *maximum* of the membership values of these elements of the fuzzy set A should be chosen as the membership of the desired output. In the other case if no element x in X is mapped to the output, then the membership value of the set B at the output is zero.

Example 2.7. Suppose that $f(x) = ax + b$ and $a \in A = \{1, 2, 3\}$ and $b \in B = \{2, 3, 5\}$ with $x = 6$. Then $f(x) = 6A + B = \{8, 15, 23\}$. □

Example 2.8. Consider the following function $y = F(s) = -2s^2 + 1$ with domain $S = R$ and range $Y = (-\infty, 1]$. Suppose that $S_f = [0, 2]$ is a fuzzy subset with the

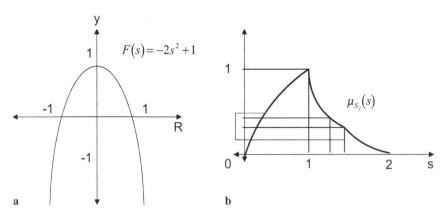

Fig. 2.14a,b Extension principle example. **a** Function: $F(s) = -2s^2 + 1$. **b** Fuzzy membership function of function $F(s) = -2s^2 + 1$

membership function shown in Fig. 2.14. The fuzzy subset $Y_f = F(S_f)$ is given by (2.10):

$$Y_f = F(S_f) = F([0, 2]) = -2[0, 2] \cdot [0, 2] + 1$$
$$= -2[0, 4] + 1 = [-8, 0]$$
$$= [-7, 1] . \tag{2.10}$$

The membership function $\mu_{Y_f}(s)$ associated with Y_f is determined as follows. Let y run through from -7 to 1. For each y, find the corresponding $s \in S_f$ satisfying $y = F(s)$, then $\mu_{Y_f}(s) = \sup_{s:F(s)=y} \mu_{S_f}(s)$. It is clear that for any $y \in [-7, 1]$, there is always one $s \in [0, 2]$ satisfying $y = F(s) = -2s^2 + 1$. Therefore, it can be easily verified that the membership function is the one shown in Fig. 2.15. □

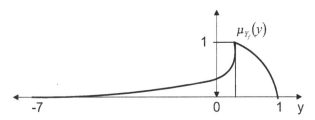

Fig. 2.15 Resulting function when the extension principle is applied

2.4.7 Alpha Cuts

An alpha cut (α-*cut*) is a crisp set of elements of A belonging to the fuzzy set to a degree α. The α-*cut* of a fuzzy set A is the crisp set comprised of all elements x of universe X for which the membership function of A is greater or equal to α (2.11):

$$A_\alpha = \{x \in X \mid \mu_A(x) \geq \alpha\} , \tag{2.11}$$

where α is in the range of $0 < \alpha \leq 1$ and "|" stands for "such that."

Example 2.9. A triangular membership function with an α-*cut* at 0.4 is shown in Fig. 2.16. Figure 2.17 shows the block diagram. □

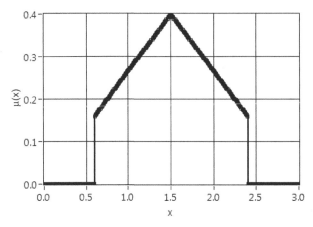

Fig. 2.16 Triangular membership function with alpha cut of 0.4

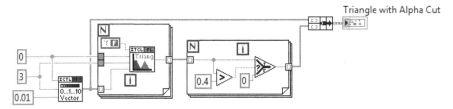

Fig. 2.17 Block diagram of the triangular membership function with alpha cut of 0.4

2.4.8 The Resolution Principle

This principle offers a way of representing membership to fuzzy sets by means of α-cuts as (2.12) denotes:

$$\mu_A(x) = \bigvee_{0 < \alpha \leq 1} [\alpha \cdot \mu_{A_\alpha}(x)] . \qquad (2.12)$$

The maximum is taken over all α-cut, and the equation indicates that the membership function of A is the union of all α-cut, after each one of them has been multiplied by α. Figure 2.18 shows these functions.

2.4.9 Fuzziness of Uncertainty

Many kinds of uncertainties arise in the real world and there are many techniques to model them. Randomness is one kind, which is typically modeled using probability theory. Outcomes are assumed to be observations of random variables and these variables have distribution laws. Fuzziness manipulates uncertainty by dealing with

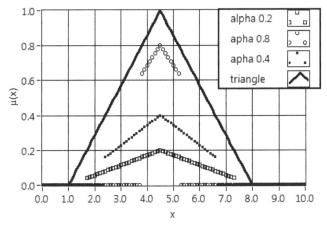

Fig. 2.18 A triangular function composed of multiple alpha cuts

the boundaries of a set that are not clearly defined. The membership in such classes is a matter of degree rather than certainty specified by fuzzy sets.

2.4.10 Possibility and Probability Theories

Possibility theory emphasizes the quantification of the semantic or meaning rather than the measure of information. The theory of possibility is analogous and yet conceptually different from the theory of probability. Probability is a measure of frequency of occurrence of an event, which has a physical event basis. Thus, probabilities have a physical event basis and are related to statistical experiments; they are primarily used for quantifying how frequently a sample occurs in a population.

Possibility theory attempts to quantify how accurately a sample resembles a *stereotype* element of a population. This stereotype is a prototypical class of the population and is known as a fuzzy set. This theory focuses more on the *imprecision* intrinsic in the language, while probability theory focuses more on the *uncertainty* of events, in the sense of its randomness in nature.

Probabilistic methods have been the instrument for quantifying equipment and human reliability as well, in which two concepts are very important: the failure rate and the error rate. Knowing these concepts and being able to control them leads to the correct understanding and function of machines, which allows industries to save money. But the correct estimation of these parameters requires a large amount of data, thus in practice they are estimated by experts based on their engineering judgment. Here is where fuzzy probabilities and possibilities can be used to model these judgments.

Over the years a new concept has emerged: the possibility theory. It is known as a fuzzy measure which is a function assigning a value between 0 and 1 to each crisp set of the universe of discourse that signifies the degree to which a particular element belongs to a set. Sugeno introduced this concept in 1974 as part of his Ph.D. dissertation.

Possibility measures are softer than probability measures, and their interpretation is quite different. On the one hand, probability is used to quantify the frequency of occurrence of an event, while on the other hand, possibility is used to quantify the meaning of an event. Possibility is an upper bound of probability, i.e., a higher degree of possibility does not imply a higher degree of probability. But if an event is not possible, then it is not probable.

If we attempt to use both probability and possibility theories to describe a similar thing, we can use the *possibility/probability consistency principle* as a guide. It will help us draw the difference between the *objectivistic* use of probability measures and *subjectivist* use of possibility or fuzzy measures.

2.5 Fuzzy Logic Theory

2.5.1 From Classical to Fuzzy Logic

Logic refers to the study of methods and principles of human reasoning. Classical logic deals with propositions that are either *true* or *false*, where each proposition has an opposite. Thus, classical logic deals with combinations of variables that represent propositions. As each variable stands for a hypothetical proposition any combination eventually assumes a truth value (true or false), but never in between the two.

The main content of classical logic is the study of *rules* that allow new logical variables to be produced as *functions* of certain existing variables. An example of a rule is shown in (2.13):

$$IF\ x_1\ is\ true\ AND\ x_2\ is\ false\ AND\ \ldots\ AND\ x_n\ is\ false\ THEN\ y\ is\ false\ . \quad (2.13)$$

The fundamental assumption upon which the classical logic is based is that every proposition is either true or false. Now it is well understood that many propositions are both partially true and false. Multi-valued logic was a first attempt to extend and generalize classical logic. In the 1930s an n-valued logic was invented by Lukasiewicz [5], which even allowed for $n = \infty$. More recently, it has been understood that there exists an isomorphism between classical logic and crisp set theory and similarly between Lukasiewicz and the fuzzy set theory.

2.5.2 Fuzzy Logic and Approximate Reasoning

The ultimate goal of fuzzy logic is to provide foundations for approximate reasoning using imprecise propositions based on fuzzy set theory, similar to classical reasoning using precise propositions based on classical set theory. We will first recall how classical reasoning works, where the following syllogism is an example of such reasoning in linguistic terms:

1. Everyone who is 70 years old is old.
2. Hiram is 70 years old and Miriam is 39 years old.
3. Hiram is old but Miriam is not.

This is an example of a precise deductive inference that is correct in the sense of classical two-valued logic. When the output variable represented by a logical formula is always true regardless of the truth values of the inputs, it is called a *tautology*. If it is the contrary then it is called a *contradiction*. Various tautologies can be used for making deductive inferences, and are referred to as *inference rules*. They can also be expressed with truth tables. The four most frequent are:

- *Modus ponens:* $(a \wedge (a \Rightarrow b)) \Rightarrow b$
- *Modus tollens:* $\left(\overline{b} \wedge (a \Rightarrow b) \right) \Rightarrow \overline{a}$
- *Syllogism:* $(a \Rightarrow b) \wedge (b \Rightarrow c) \Rightarrow (a \Rightarrow c)$
- *Contraposition:* $(a \Rightarrow b) \Rightarrow \left(\overline{b} \Rightarrow \overline{a} \right)$.

We will now consider an example of approximate reasoning in linguistic terms that cannot be handled by the classical reasoning logic:

1. Everyone who is 60 to 70 years old is old, but very old if 71 years old or above; everyone who is 20 to 39 is young but very young if 19 years old or below.
2. Hiram is 70 years old and Miriam is 39 years old.
3. Hiram is old but not very old; Miriam is young but not very young.

This is an example of *approximate reasoning*; in order to deal with an imprecise inference, fuzzy logic can be employed. It allows imprecise linguistic terms such as:

- Fuzzy predicates: rare, expensive, fast, high
- Fuzzy quantifiers: few, usually, much, little
- Fuzzy truth values: true, unlikely true, false, and mostly false.

To describe fuzzy logic mathematically, the following concepts and notations are introduced. Let **S** be a universe set and A a fuzzy set associated with a membership function $\mu_A(x)$, $x \in S$. If $y = \mu_A(x_0)$ is a point in $[0, 1]$, representing the truth value of the proposition "x_0 is a," then the value for "x_0 is not a" is:

$$\overline{y} = \mu_A(x_0 \text{ is not } a) = 1 - \mu_A(x_0 \text{ is } a) = 1 - \mu_A(x_0) = 1 - y. \quad (2.14)$$

Consequently for n members $x_1, \ldots, x_n$ in **S** with n corresponding truth values $y_i = \mu_A(x_i)$ in $[0, 1]$, $i = 1, \ldots, n$ by applying the extension principle, the truth values of "not a" get defined as $\overline{y}_i = 1 - y_i$, $i = 1, \ldots, n$. We note that $n = \infty$ is allowed. The same can be applied to other logical operators.

For instance in the modus ponens $(a \wedge (a \Rightarrow b)) \Rightarrow b$, the inference rule is: *IF* $\mu_A(a) > 0$ *AND* $\mu_A(a \Rightarrow b) = \min\{1, 1 + \mu(b) - \mu(a)\} > 0$ *THEN*

$\mu_B(a) > 0$, where $\mu > 0$ is equivalent to $\mu \in (0, 1]$. This fuzzy logic inference can be interpreted as follows: *IF a* is true with a certain degree of confidence *THEN b* is true with a certain degree of confidence. All of these degrees of confidence can be quantitatively evaluated by using the corresponding membership functions. This is a *generalized modus ponens*, called *fuzzy modus ponens*.

2.5.3 Fuzzy Relations

In fuzzy relations we consider n-tuples of elements that are related to a *degree*. Just as the question of whether some element belongs to a set may be considered a matter of degree, whether some elements are associated may also be a matter of degree. Fuzzy relations are fuzzy sets defined on Cartesian products. While fuzzy sets are defined on a single universe of discourse, fuzzy relations are defined on higher-dimensional universes of discourse.

If **S** is the universe set and A and B are subsets, $A \times B$ will denote a product set in the universe $\mathbf{S} \times \mathbf{S}$. A fuzzy relation is a relation between elements of A and elements of B, described by a membership function $\mu_{A \times B}(a, b)$, $a \in A$ and $b \in B$.

A discrete example of a fuzzy relation can be defined as: $\mathbf{S} = R$, $A = \{a_1, a_2, a_3, a_4\} = \{1, 2, 3, 4\}$ and $B = \{b_1, b_2, b_3\} = \{0, 0.1, 2\}$. Table 2.5 defines a fuzzy relation: a is considerably larger than b.

Table 2.5 Definition of relation: a is considerably larger than b

	b_1	b_2	b_3
a_1	0.6	0.6	0.0
a_2	0.8	0.7	0.0
a_3	0.9	0.8	0.4
a_4	1.0	0.9	0.5

2.5.4 Properties of Relations

Fuzzy relations can be represented in many ways: linguistically, listed, in a directed graph, tabular, matrix, among others. Crisp and fuzzy relations are classified on the basis of the mathematical properties they possess. In fuzzy relations, different properties call for different requirements for the membership function of a relation. The following are some of the properties that a relation can have:

- *Reflexive*. We say that a relation R is *reflexive* if any arbitrary element x in S for which xRx is *valid*.

- *Anti-reflexive.* A relation R is *anti-reflexive* if there is no x in S for which xRx is valid.
- *Symmetric.* A relation R is *symmetric* if for all x and y in S, the following is true: if xRy then yRx is valid also.
- *Anti-symmetric.* A relation R is *anti-symmetric* if for all x and y in S, when xRy is valid and yRx is also valid, then $x = y$.
- *Transitive.* A relation R is called *transitive* if the following for all x, y, z in S: if xRy is valid and yRx is also valid, then xRz is valid as well.
- *Connected.* A relation R is *connected* when for all x, y in S, the following is true: if $x \neq y$, then either xRy is valid or yRx is valid.
- *Left unique.* A relation R is *left unique* when for all x, y, z in S the following is true: if xRy is valid and yRx is also valid, then we can infer that $x = y$.
- *Right unique.* A relation R is *right unique* when for all x, y, z in S the following is true: if xRy is valid and xRz is also valid, then we can infer that $y = z$.
- *Biunique.* A relation R that is both *left unique* and *right unique* is called *biunique*.

2.5.5 Max–Min Composition

Let R, R^1, R^2, R^3 be fuzzy relations defined on the same product set $A \times A$ and let $\circ$ be the max–min composition operation for these fuzzy relations. Then:

1. The max–min composition is *associative* (2.15):

$$\left(R^1 \circ R^2\right) \circ R^3 = R^1 \circ \left(R^2 \circ R^3\right) . \qquad (2.15)$$

2. If R^1 is reflexive and is arbitrary R^2 is arbitrary, then $\mu_{R^2}(a, b) \leq \mu_{R^1 \circ R^2}(a, b)$ for all $a, b \in A$ and $\mu_{R^2}(a, b) \leq \mu_{R^2 \circ R^1}(a, b)$ for all $a, b \in A$.
3. If R^1 and R^2 are reflexive, then so are $R^1 \circ R^2$ and $R^2 \circ R^1$.
4. If R^1 and R^2 are symmetric and $R^1 \circ R^2 = R^2 \circ R^1$, then $R^1 \circ R^2$ is symmetric. In particular if R is symmetric then so is $R \circ R$.
5. If R is symmetric and transitive, then $\mu_R(a, b) \leq \mu_R(a, a)$ for all $a, b \in A$.
6. If R is reflexive and transitive, then: $R \circ R = R$.
7. If R^1 and R^2 are transitive and $R^1 \circ R^2 = R^2 \circ R^1$, then $R^1 \circ R^2$ is transitive.

An example will show how the max–min composition works for the approximate reasoning.

Example 2.10. Supposing that we have two relations R^1 and R^2, we want to compute the max–min composition of the two, $R = R^1 \circ R^2$. The relationships to be composed are described in Tables 2.6 and 2.7.   □

To find the new relation we use the definition of the max–min composition: $\mu_{R_1 \circ R_2}(x, z) = \bigvee_y [\mu_{R_1}(x, y) \wedge \mu_{R_2}(y, z)]$. To make the composition we proceed in the following manner. First, we fix x and z and vary y. Next we evaluate the following

Table 2.6 Definition of relation: R^1

R^1	y_1	y_2	y_3	y_4	y_5
x_1	0.1	0.2	0.0	1.0	0.7
x_2	0.3	0.5	0.0	0.2	1.0
x_3	0.8	0.0	1.0	0.4	0.3

Table 2.7 Definition of relation: R^2

R^2	y_1	y_2	y_3	y_4
x_1	0.9	0.0	0.3	0.4
x_2	0.2	1.0	0.8	0.0
x_3	0.8	0.0	0.7	1.0
x_4	0.4	0.2	0.3	0.0
x_5	0.0	1.0	0.0	0.8

pairs of minima, as shown in (2.16):

$$\mu_{R_1}(x_1, y_1) \wedge \mu_{R_2}(y_1, z_1) = 0.1 \wedge 0.9 = 0.1$$
$$\vdots$$
$$\mu_{R_1}(x_1, y_5) \wedge \mu_{R_2}(y_5, z_1) = 0.7 \wedge 0.0 = 0.0$$

(2.16)

We take the maximum of these terms and obtain the value of the (x_1, z_1) element of the relation as in: $\mu_{R_1 \circ R_2}(x_1, z_1) = 0.1 \vee 0.2 \vee 0.0 \vee 0.4 = 0.4$. We then determine the grades of membership for all other pairs and we finally obtain R as shown in Table 2.8.

Table 2.8 Definition of min–max composition: R

$R = R^1 \circ R^2$	z_1	z_2	z_3	z_4
x_1	0.4	0.7	0.3	0.7
x_2	0.3	1.0	0.5	0.8
x_3	0.8	0.3	0.7	1.0

2.5.6 Max–Star Composition

Different operations can be used in place of min in the max–min composition while still performing maximization. This type of composition is known as max–star or max-∗-composition. It is defined in (2.17). The integral sign in this equation is replaced by summation when the product is discrete.

$$R^1 * R^2 \equiv \int_{X \times Z} \bigvee_y [\mu_{R^1}(x, y) * \mu_{R^2}(y, z)]/(x, z).$$

(2.17)

2.5.7 Max–Average Composition

In a max–average composition the arithmetic sum divided by 2 is used. Thus the max–average composition of R^1 with R^2 is a new relation $R^1 \langle + \rangle R^2$ (2.18):

$$R^1 \langle + \rangle R^2 \equiv \int_{X \times Z} \bigvee_y \left[\frac{1}{2} \left(\mu_{R^1} (x, y) + \mu_{R^2} (y, z) \right) \right] / (x, z) . \qquad (2.18)$$

2.6 Fuzzy Linguistic Descriptions

Fuzzy linguistic descriptions are often called *fuzzy systems* or *linguistic descriptions*. They are formal representations of systems made through fuzzy *IF–THEN* rules. A linguistic variable is a variable whose arguments are words modeled by *fuzzy sets*, which are called *fuzzy values*. They are an alternative to analytical modeling systems. Informal linguistic descriptions used by humans in daily life as well as in the performance of skilled tasks are usually the starting point for the development of fuzzy linguistic descriptions. Although fuzzy linguistic descriptions are formulated in a human-like language, they have rigorous mathematical foundations involving fuzzy sets and relations. The knowledge is encoded in a statement of the form shown in (2.19):

IF (a set of conditions is satisfied) *THEN* (a set of consequences can be inferred).
$$\text{(2.19)}$$

A *general fuzzy IF–THEN rule* has the form:

$$IF \ a_1 \ is \ A_1 \ AND \ldots AND \ a_n \ is \ A_n \ THEN \ b \ is \ B . \qquad (2.20)$$

Using the fuzzy logic *AND* operation, this rule is implemented by:

$$\mu_A (a_1) \wedge \ldots \mu_{A_n} (a_n) \Rightarrow \mu_B (b) . \qquad (2.21)$$

Reasoning with Fuzzy Rules

Fuzzy reasoning includes two distinct parts: evaluating the rule antecedent (*IF* part of the rule) and implication or applying the result to the consequent, the *THEN* part of the rule. While in classical rule-based systems if the antecedent of the rule is true, then the consequent is also true, but in fuzzy systems the evaluation is different. In fuzzy systems the antecedent is a fuzzy statement, this means all the rules fire at some extent. If the antecedent is true in some degree of membership, then the consequent is also true in some degree.

Example 2.11. Consider two fuzzy sets, *tall men* and *heavy men* represented in Fig. 2.19. These fuzzy sets provide the basis for a weight estimation model. The model is based on a relationship between a man's height and his weight, which can be expressed with the following rule: *IF* height is *tall*, *THEN* weight is *heavy*. The value of the output or the membership grade of the rule consequent can be estimated directly from a corresponding membership grade in the antecedent. Fuzzy rules can have multiple antecedents, as the consequent of the rule, which can also include multiple parts.   □

In general, fuzzy expert systems incorporate not one but several rules that describe expert knowledge. The output of each rule is a fuzzy set, but usually we need to obtain a single number representing the expert system output, the crisp solution. To obtain a single crisp output a fuzzy expert system first aggregates all output fuzzy sets into a single output fuzzy set, and then defuzzifies the resulting set into a single number.

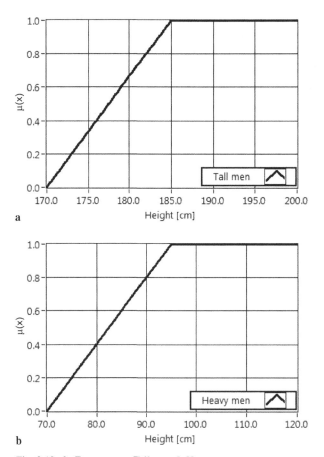

Fig. 2.19a,b Fuzzy sets. **a** Tall men. **b** Heavy men

2.7 The Fuzzy Logic Controller

With traditional sets an element either belongs to the set or does not belong to the set {0,1}, while in fuzzy sets the degree to which the element belongs to the set is analyzed and it is called the membership degree, giving values in the range [0,1], where 1 indicates that the element belongs completely to the set. The *fuzzy logic controllers* (FLC) make a non-linear mapping between the input and the output using *membership functions* and *linguistic rules* (normally in the form if__then__).

In order to use a FLC, knowledge is needed and this can be represented as two different types:

1. *Objective* information is what can be somehow quantifiable by mathematical models and equations.
2. *Subjective* information is represented with linguistic rules and design requirements.

2.7.1 Linguistic Variables

Just like in human thinking, in fuzzy logic systems (FLS) linguistic variables are utilized to give a "value" to the element, some examples are *much, tall, cold*, etc. FLS require the linguistic variables in relation to their numeric values, their quantification and the connections between variables and the possible implications.

2.7.2 Membership Functions

In FLS the membership functions are utilized to find the degree of membership of the element in a given set.

2.7.3 Rules Evaluation

The rules used in the FLS are of the *IF–THEN* type, for example, *IF* x_1 is big *THEN* y_1 is small. To define the rules you need an expert, or you must be able to extract the information from a mathematic formula. The main elements of a FLC are fuzzification, rules evaluation, and defuzzification, as shown in Fig. 2.20.

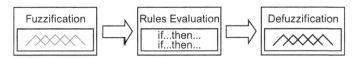

Fig. 2.20 Elements of the FLC

2.7.4 Mamdani Fuzzy Controller

The Mamdani is one kind of fuzzy controller. This section gives an introduction to the Mamdani fuzzy controller.

2.7.5 Structure

This controller consists of three main parts: fuzzification, rules evaluation and defuzzification. The inputs have to be crisp values in order to allow the fuzzification using membership functions, and the outputs of this controller are also crisp values. This controller is shown in Fig. 2.21.

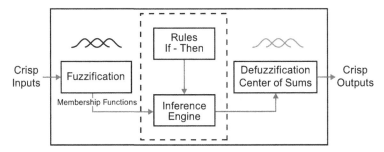

Fig. 2.21 Mamdani block diagram

2.7.6 Fuzzification

For mapping the crisp values to fuzzy ones, you have to evaluate their membership degree using membership functions. With this you get one fuzzy value for each crisp input.

An example is presented in Fig. 2.22 where $\mu(a)$ is the membership value, and the crisp values are a_0.

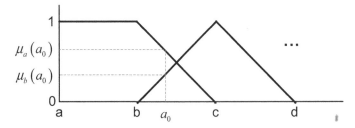

Fig. 2.22 Membership functions (input)

2.7.7 Rules Evaluation

After getting the membership values of the inputs, they are evaluated using *IF–THEN* rules: *IF a* is *x AND b* is *y AND c* is *z THEN w*, where *a*, *b* and *c* are the crisp inputs, *x*, *y* and *z* are the fuzzy clusters to which the inputs may correspond, and *w* is the output fuzzy cluster used to defuzzify. To be able to obtain the fuzzy values of the outputs, the system has to use an inference engine. The min–max composition is used which takes the minimum of the premises and the maximum of the consequences.

2.7.8 Defuzzification

To defuzzify the outputs we use the center of sums method. In this method we take the output from each contributing rule, and then we add them. The center of sums is one of the most popular methods for defuzzificating because it is very easy to implement and gives good results. With (2.22) we get the crisp value of the outputs, and Fig. 2.23 shows the graphical discrete representation.

$$u^* = \frac{\sum_{i=1}^{N} u_i \cdot \sum_{i=1}^{N} \mu_{A_k}(u_i)}{\sum_{i=1}^{N} \sum_{k=1}^{n} \mu_{A_k}(u_i)} . \tag{2.22}$$

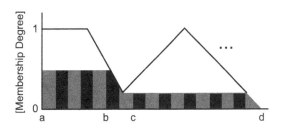

Fig. 2.23 Membership functions (output)

2.7.9 Tsukamoto Fuzzy Controller

Tsukamoto controllers are like the Mamdani controllers but with monotonic input and output membership functions. A monotonic function is a function that preserves the given order. In other words, it increases or decreases, and if its first derivative (needs not be continuous) does not change in sign.

2.7.10 Takagi–Sugeno Fuzzy Controller

The Takagi–Sugeno is another kind of fuzzy controller; the following gives an introduction to this kind of controller. The main difference in Sugeno type is in the defuzzificatio stage. In this case we do not use membership functions anymore.

2.7.11 Structure

This controller consists of three main parts: fuzzification, rules evaluation and defuzzification (see Fig. 2.24). The inputs have to be crisp values in order to allow the fuzzification to use membership functions, and the outputs of this controller are also crisp values.

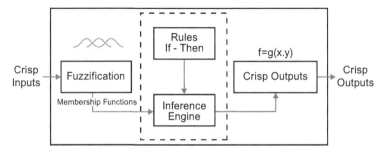

Fig. 2.24 Takagi–Sugeno diagram

2.7.12 Fuzzification

In order to transform the crisp values into fuzzy ones, you have to evaluate those using membership functions. With this you get one fuzzy value for each crisp input. The membership functions could be either conventional ones or non-conventional, the selection of the membership function depends on the specific problem. As result the first step in fuzzy logic is to select the best membership function for describing the problem.

2.7.13 Rules Evaluation

After getting the fuzzy values of the inputs, they are evaluated using *IF–THEN* rules: *IF a is x AND b is y AND c is z THEN u = f(a,b,c)*, where *a*, *b* and *c* are the crisp inputs, x, y and z are the fuzzy clusters to which the inputs could correspond, and $u = f(a,b,c)$ is a polynomial. Instead of evaluating a, b and c, we evaluate in the polynomial the number of the fired rules.

These polynomials are calculated using regressions in order to adjust them to the desired output functions. For the inference system we use the minimum of the premises, but because the consequence of the rule is not fuzzy, the maximum of the consequences could not be used.

2.7.14 Crisp Outputs

To defuzzify take the product of the sum of the minimum of the antecedents of every rule fired times the value of the polynomial evaluated in that fired rule, all this divided by the sum of the minimum of the antecedents of every rule fired:

$$\text{Output} = \frac{\sum_{i-1}^{r} [\min(\mu_{ix,y,z})u(a,b,c)]}{\sum_{i-1}^{r} \min(\mu_{ix,y,z})}. \tag{2.23}$$

2.8 Implementation of the Fuzzy Logic Controllers Using the Intelligent Control Toolkit for LabVIEW

We will now create a FLC using the Intelligent Control Toolkit for LabVIEW (ICTL). The fuzzy logic VI includes the following parts:

- Mamdani controller.
- Takagi–Sugeno controller.
- Tsukamoto controller.

It is important to state that there are no limitations on the number of linguistic variables and linguistic terms, nor on the type of membership functions supported. For a better understanding of how to use the FLC VIs, the implementation of FLCs in a mobile robot will be explained step-by-step in the following.

A robot named Wheel sends via Bluetooth® the distance measured by right, center, and left ultrasonic distance sensors. It has two servomotors that receive 0 for no movement, 1 to go clockwise, and 2 for counterclockwise, as in the diagram in Fig. 2.25.

The robot is driven by a BASIC Stamp® controller, the Bluetooth antenna is an EmbeddedBlue™ Transceiver from Parallax, and the two servos and three Ping)))™ sensors are also from Parallax. For the computer we use a Belkin Bluetooth antenna. Figure 2.26 plots the information flow.

Fig. 2.25 Three-wheeled
robot parts

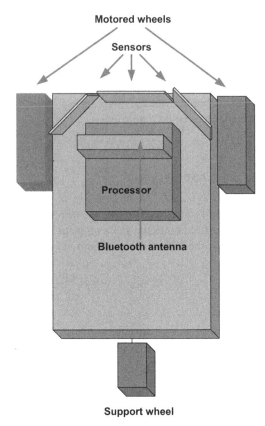

Fig. 2.26 Three-wheeled
robot diagram

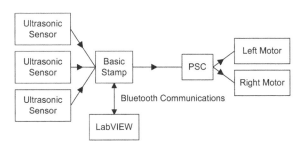

2.8.1 Fuzzification

We know that the controller for the robot receives three inputs (left, center, and right) and generates two outputs (sLeft and sRight). With this information we can calculate the number of rules, which is $3^2 = 9$.

Array-Bell-Fu... Array-Tria-Fu... Bell-Function.vi Bell-multi-inpu... Monotonic_lin...

Monotonic_lo... Tria-multi-inp... triang-functio... triangular-fun...

Fig. 2.27 Function VI blocks

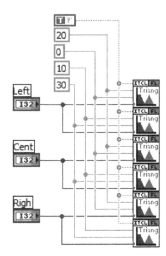

Fig. 2.28 Block diagram of
the membership function

Next thing we need to do is generate the fuzzy membership functions for the fuzzification of the inputs; as stated before, we will be using two membership functions with the linguistic values of *close* and *far*. We can choose from different forms for the functions in the fuzzification process, as shown in Fig. 2.27.

For our controller we will choose the triangular function, called **triang-function. vi**, because we will only have two membership functions per input (Fig. 2.28). We will use them as shoulder triangular functions, and because our robot is small and fast we will set the *close* cluster within the limits of 0 to 20 cm, and the *far* cluster limits will be 10 to 30 cm. Because of the programming of the triangular function, the decreasing or increasing of the functions will begin just between the limit where the membership value starts to be zero a and the final limit where it starts to be one b. If $a < b$ the function will be a left shoulder function, otherwise it will be a right shoulder.

Now we need to create a 2D array from the results obtained from the fuzzification process. Figure 2.29 shows the wiring part of the code, and Fig. 2.30 shows the wiring process. Basically for each input an array with their values must be created and then a second array must be created containing all the input information.

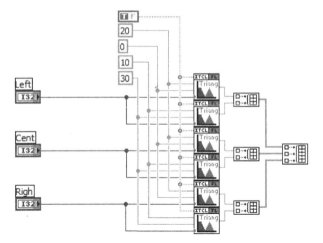

Fig. 2.29 Complete block diagram of the membership function

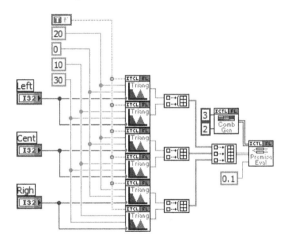

Fig. 2.30 Block diagram until
rules evaluation

2.8.2 Rules Evaluation

With this we will have the information of the membership value for *left* in the row
of the 2D array, the second row will have the information for *center*, and the third
array will have the information for the *right* array. For each row, the first value will
be for the *close* set, and the second value will be for the *far* set (Table 2.9). This
will be pretty much like creating a table for the evaluation of the inputs. There is
a VI that will automatically generate this table for us: it is the input combinatory
generator and we must feed it the number of inputs and membership functions.

This way we have generated the premise part of the rule: *IF left* is *close AND center*
is *close AND right* is *close, THEN*. . . Now we need to use the **premise_evaluation.vi**
to obtain the min operation for each rule. Sometimes we will want to consider that
a rule is activated only if its min value is above a certain level; that is why we have

Table 2.9 Rules base

Input from the sensors		
Left	Center	Right
Close	Close	Close
Close	Close	Far
Close	Far	Close
Close	Far	Far
Far	Close	Close
Far	Close	Far
Far	Far	Close
Far	Far	Far

set the "value" input from the premise VI to 0.1. This part of the code is shown in Fig. 2.30.

Now we need to generate the combinations that will make the robot move around. We have previously established that 2 = *move forward*, 1 = *move backward*, and 0 = *stopped*. With this in mind we can generate a table with the consequences of the rules. In order to obtain the values we must generate 1D arrays for the defuzzification of each output (Table 2.10).

Table 2.10 Rules of outputs for the Takagi–Sugeno and Tsukamoto controllers

Output for the wheels		
Action	Left wheel	Right wheel
Go backward	2	2
Turn right back	2	0
Go front	1	1
Turn right	2	1
Turn left back	0	2
Go backward	2	2
Turn left	1	2
Go front	1	1

2.8.3 Defuzzification: Crisp Outputs

In order to obtain the crisp outputs for a controller using singleton output functions or a Takagi–Sugeno inference, we must use the **defuzzifier_constants.vi** that will help us obtain the final outputs sLeft and sRight. In case we would like to use equations instead of constants, we would need to evaluate these equations and pass the constant value to the **defuzzifier_constants.vi**; this way we are able to use any kind of equation. Figure 2.31 shows the complete block diagram of Takagi

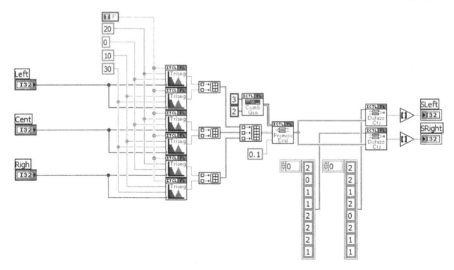

Fig. 2.31 Complete block diagram of the Takagi–Sugeno and Tsukamoto controllers

Table 2.11 Rules set for the Takagi–Sugeno and Tsukamoto controllers

Input from the sensors			Output for the wheels		
Left	Center	Right	Action	Left wheel	Right wheel
Close	Close	Close	Go backward	2	2
Close	Close	Far	Turn right back	2	0
Close	Far	Close	Go front	1	1
Close	Far	Far	Turn right	2	1
Far	Close	Close	Turn left back	0	2
Far	Close	Far	Go backward	2	2
Far	Far	Close	Turn left	1	2
Far	Far	Far	Go front	1	1

and Tsukamoto controllers. This way we have created a controller that will behave according to the set of rules shown in Table 2.11.

If we decide to create a Mamdani controller, the premise part is pretty much the same, only the consequences part will have to change, i.e., instead of using singletons we will use three triangular functions for the defuzzification process called *stopped*, *forward*, and *backward*. We have to use the **general_defuzzifier_mamdani.vi** for each output that we would like to generate. This VI will take the number of membership functions that will be used for defuzzification, the same set created with the desired output for each fired rule and the result of the evaluation of the premises, and it will return the max for each of the desired sets. With this information we will then have to create the defuzzifying functions and pass them their respective max value. Finally we have to combine the results to obtain the desired outputs. Figure 2.32 shows the complete block diagram of the Mamdani controller. This controller will behave according to the set of rules shown in Table 2.12.

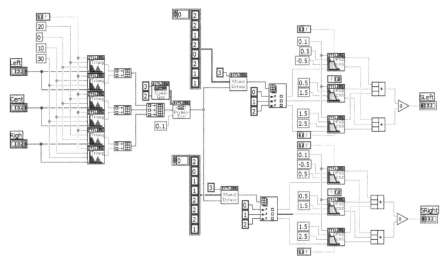

Fig. 2.32 Complete block diagram of a Mamdani controller

Table 2.12 Rules set for the Mamdani controller

Input from the sensors			Output for the wheels		
Left	Center	Right	Action	Left wheel	Right wheel
Close	Close	Close	Go backward	Backward	Backward
Close	Close	Far	Turn right back	Backward	Stopped
Close	Far	Close	Go front	Forward	Forward
Close	Far	Far	Turn right	Backward	Forward
Far	Close	Close	Turn left back	Stopped	Backward
Far	Close	Far	Go backward	Backward	Backward
Far	Far	Close	Turn left	Forward	Backward
Far	Far	Far	Go front	Forward	Forward

2.9 Classical Control Example

Here fuzzy logic is used to control the speed of the discrete model of a direct current motor. A fuzzy proportional derivative controller was used along with the model of a direct current motor whose transfer function in continuous time is shown in (2.24):

$$G\left(s\right) = \frac{2.445}{s^2 + 6.846s}.\qquad(2.24)$$

The model of the direct current motor was digitalized with a period of 0.1 s and converted to difference equations, then programmed as shown in Fig. 2.33. The diagram for the closed-loop controller is shown in Fig. 2.34.

The error and the difference of the error is calculated and sent to the fuzzy controller that will create the appropriate signal to control the speed reference of the motor. The motor will then react to the control signals of the controller and the speed will be measured and receive feedback.

Fig. 2.33 Digital model of the plant in difference equation form

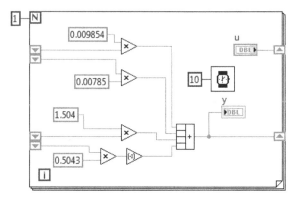

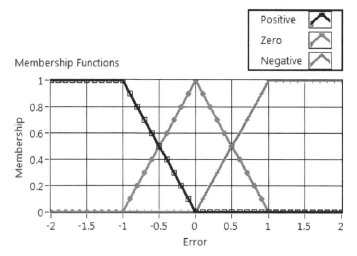

Fig. 2.34 Closed-loop diagram of the controller

Membership Functions

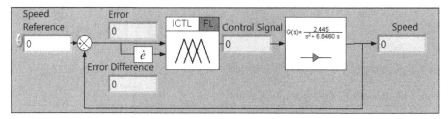

Fig. 2.35 Membership functions for fuzzification and defuzzification

The proportional derivative fuzzy controller is a Mamdani controller with two inputs and one output. Three membership functions for fuzzification and defuzzification are used of triangular form, as shown in Fig. 2.35. The diagram of the fuzzy controller is shown and explained in Fig. 2.36.

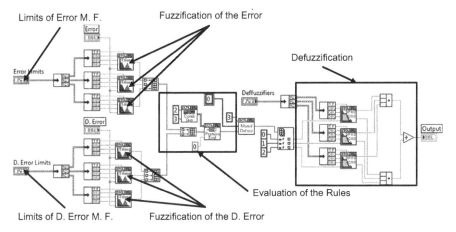

Fig. 2.36 Block diagram of fuzzy controller

The fuzzy controller works like this:

1. The limits of the membership functions are set.
2. Then the error and the difference error are evaluated in three triangular membership functions; their outputs are then gathered in an array to be used.
3. After that the process of evaluation of the rules is complete.
4. Finally three triangular membership functions are used to defuzzify the output of the controller.

The code for the complete program is shown in Fig. 2.37.

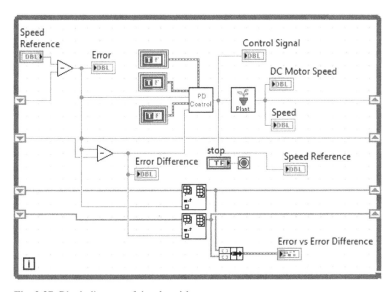

Fig. 2.37 Block diagram of the closed-loop

References

1. Zadeh LA (1973) Outline of a new approach to the analysis of complex systems and decision processes, IEEE Trans Syst Man Cyber SMC 3:28–44
2. Karray FO, de Silva C (2004) Soft computing and intelligent system design. Addison Wesley Longman, Boston
3. von Altrock C (1994) Fuzzy logic technologies in automotive engineering. IEEE Proceedings of WESCON, Anaheim, CA, 27–29 Sept 1994
4. Zadeh LA (1965) Fuzzy sets. Info Control 8(3):338–353
5. Tsoukalas LH, Uhrig RE (1996) Fuzzy and neural approaches in engineering. Wiley, New York

Futher Reading

Guanrong C, Trung P (2000) Introduction to fuzzy sets, fuzzy logic, and fuzzy control systems. CRC, Boca Raton, FL

Hung NT, Elbert W (1999) A first course in fuzzy logic, 2nd edn. CRC, Boca Raton, FL

Timothy RJ (1995) Fuzzy logic with engineering applications. McGraw-Hill, Boston

Chapter 3
Artificial Neural Networks

3.1 Introduction

The human brain is like an information processing machine. Information can be seen as signals coming from senses, which then runs through the nervous system. The brain consists of around 100 to 500 billion neurons. These neurons form clusters and networks. Depending on their targets, neurons can be organized hierarchically or layered.

In this chapter we present the basic model of the neuron in order to understand how neural networks work. Then, we offer the classification of these models by their structures and their learning procedure. Finally, we describe several neural models with their respective learning procedures. Some examples are given throughout the chapter.

Biologically, we can find two types of cell groups in the nervous system: the *glial* and *nervous cells*. The nervous cells, also called neurons, are organized in a functional *syncytium* way. This is a complex distribution, which can be imagined as a computer network or a telephone system network. Neurons communicate through an area called a *synapse*. This region is a contact point of two neurons.

The main components of the neurons are the *axon* and the *dendrite*. Each neuron has only one axon, which is divided into multiple exits on a later stage. The number of dendrites may vary and each of them is an extension of the neuron's body, which increases the number of entries of information collection. The axon is the only exit point of the neuron and can be up to 120 mm long.

In a simplified way, a neuron functions in a way where none, one or many electrical impulses are received through its dendrites from axons from other neurons. Those electrical impulses are added in order to have a final potential. This potential must exceed a certain level to have the neuron generate an electrical impulse on its axon. If the level required is not met, then the axon of that neuron does not fire its axon.

In other words, neurons can be divided into *dendrites*, which are channels of input signals, a *core cell* that processes all these signals, and *axons* that transmit output

Fig. 3.1 Schematic drawing
of a biological neuron

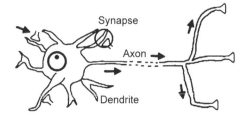

signals of the processed information from dendrites. Figure 3.1 shows a schematic drawing of a neuron.

Why would we want to look inside neurons? The human brain processes information and can react from distinct stimuli. Moreover, the brain can generalize this information to act when new situations are presented. If we are looking inside neurons, we are in a sense searching the notion of how the human brain learns and generalizes information.

At this point, neurons can be modeled as follows. Dendrites are connected to some axons from other neurons, but some links are reinforced when typical actions occur. However, links are not very strong when these channels are not used. Therefore, input signals are weighted by this reinforcement (positively or negatively). All these signals are then summarized and the core cell processes that information. This process is modeled mathematically by an *activation function*. Finally, the result is transmitted by axons and the output signal of the neuron goes to other cells. Figure 3.2 shows this neural model.

The activation function is the characterization of the neurons' activities when input signals stimulate them. Then, the activation function can be any kind of function that describes the neural processes. However, the most common are sigmoidal func-

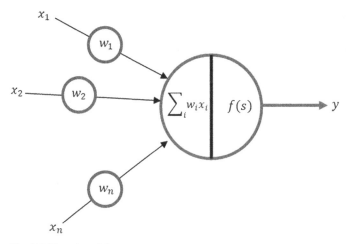

Fig. 3.2 Neural model

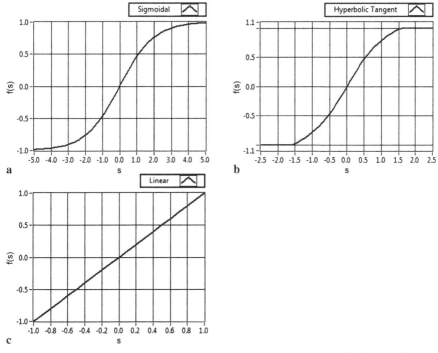

Fig. 3.3a–c Graphs of typical activation functions. **a** Sigmoidal function: $f_s(x) = \frac{2}{e^x+1} - 1$. **b** Hyperbolic tangent function: $f_{\text{tanh}}(x) = \tanh(x)$. **c** Linear function: $f_{\text{linear}}(x) = x$

tion, linear function, and hyperbolic tangent function. Figure 3.3 shows the graphical form of these functions.

These functions can be implemented in LabVIEW with the Intelligent Control Toolkit in the following way. First, open the ICTL and select the *ANN* (artificial neural network) section. We can find all the possibilities in ANN as seen in Figs. 3.4 and 3.5. The activation functions are in ICTL ≫ ANN ≫ Backpropagation ≫ NN

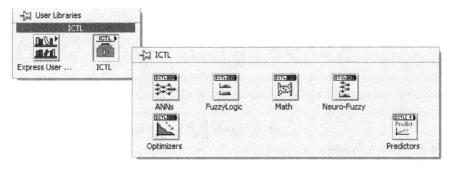

Fig. 3.4 Accessing ANNs library in the ICTL

Fig. 3.5 ANNs library in
ICTL

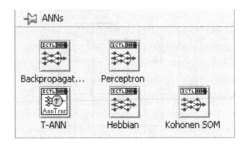

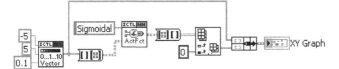

Fig. 3.6 Block diagram of the Sigmoidal function

Methods ≫ **activationFunction.vi**. For instance, we want to plot the sigmoidal
function. Then, we can create a VI like that in Fig. 3.6.

At first, we have a **rampVector.vi** that creates an array of elements with values
initialized at −5 and runs with a stepsize of 0.1 through to 5. The activation function
VI must be a real-valued matrix. This is the reason why the vector is transformed
into a matrix. Also, the output of this VI is a matrix. Then, we need to get all values
in the first column. Finally, the array in the interval [−5, 5] is plotted against values
$f(x)$ for all $x \in [-5, 5]$.

The graph resulting from this process is seen in Fig. 3.3. If we want to plot
the other function, the only thing we need to do is change the label *Sigmoidal* to
Hyperbolic Tangent in order to plot the hyperbolic tangent function, or *User Defined*
if we need a linear function. Figure 3.7 shows these labels in the block diagram.

Example 3.1. Let $X = \{0.4, -0.5, 0.2, -0.7\}$ be the input vector and $W = \{0.1, 0.6, 0.2, 0.3\}$ be the weight vector. Suppose a sigmoidal activation function is the processing of the core cell. (a) What is the value of the output signal? (b) What is the value of the output signal if we change the activation function in (3.1), known as symmetrical hard limiting?

$$f(s) = \begin{cases} -1 & s \le 0 \\ 1 & s > 0 \end{cases} \qquad (3.1)$$

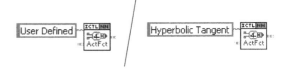

Fig. 3.7 Labels in the activation function VI

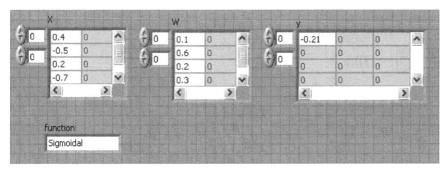

Fig. 3.8 Calculations of the output signal

Solution. (a) We need to calculate the inner product of the vector X and W. Then, the real-value is evaluated in the sigmoidal activation function.

$$y = f_{\text{sigmoidal}}\left(\sum_i w_i x_i = (0.4)(0.1) + (-0.5)(0.6) + (0.2)(0.2) + (-0.7)(0.3)\right.$$

$$\left. = -0.43\right) = -0.21 \tag{3.2}$$

This operation can be implemented in LabVIEW as follows. First, we need the NN (neural network) VI located in the path ICTL $\gg$ ANNs $\gg$ Backpropagation $\gg$ NN Methods $\gg$ **neuralNetwork.vi**. Then, we create three real-valued matrices as seen in Fig. 3.8. The block diagram is shown in Fig. 3.9. In view of this block diagram, we need some parameters that will be explained later. At the moment, we are interested in connecting the X-matrix in the *inputs* connector and **W**-matrix in the *weights* connector. The label for the activation function is *Sigmoidal* in this example but can be any other label treated before. The condition 1 in the $L - 1$ connector comes from the fact that we are mapping a neural network with four inputs to one output. Then, the number of layers L is 2 and by the condition $L - 1$ we get the number 1 in the blue square. The 1D array {4, 1} specifies the number of neurons per layer, the input layer (four) and the output layer (one). At the *globalOutputs* the y-matrix is connected.

From the previous block diagram of Fig. 3.9 mixed with the block diagram of Fig. 3.6, the connections in Fig. 3.10 give the graph of the sigmoidal function evaluated at -0.43 pictured in Fig. 3.11. Note the connection comes from the **neuralNet-**

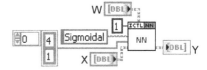

Fig. 3.9 Block diagram of Example 3.1

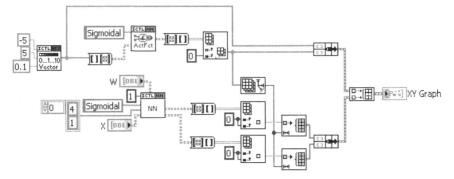

Fig. 3.10 Block diagram for plotting the graph in Fig. 3.11

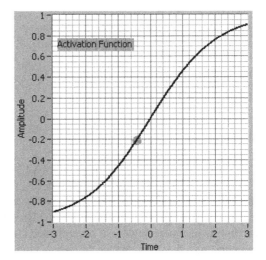

Fig. 3.11 The value −0.43
evaluated at a Sigmoidal
function

work.vi at the *sumOut* pin. Actually, this value is the inner product or the sum of the
linear combination between X and W. This real value is then evaluated at the activation function. Therefore, this is the x-coordinate of the activation function and the
y-coordinate is the *globalOutput*. Of course, these two out-connectors are in matrix
form. We need to extract the first value at the position $(0, 0)$ in these matrices. This
is the reason we use the matrix-to-array transformation and the index array nodes.
The last block is an initialize array that creates a 1D array of m elements (sizing
from any vector of the sigmoidal block diagram plot) with the value −0.43 for the
sumOut connection and the value −0.21 for the *globalOutput* link. Finally, we create an array of clusters to plot the activation function in the interval $[−5, 5]$ and the
actual value of that function.

  (b) The inner product is the same as the previous one, −0.43. Then, the activation
function is evaluated when this value is fired. So, the output value becomes −1. This
is represented in the graph in Fig. 3.12. The activation function for the symmetric
hard limiting can be accessed in the path ICTL ≫ ANNs ≫ Perceptron ≫ Trans-

Fig. 3.12 The value −0.43 evaluated at the symmetrical hard limiting activation function

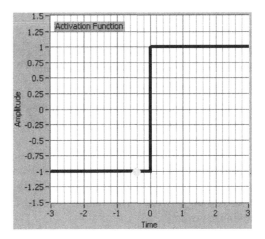

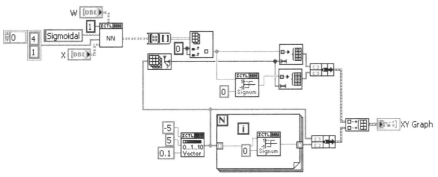

Fig. 3.13 Block diagram of the plot in Fig. 3.12

fer F. ≫ **signum.vi**. The block diagram of Fig. 3.13 shows the next explanation. In this diagram, we see the activation function below the NN VI. It consists of the array in the interval $[-5, 5]$ and inside the for-loop is the symmetric hard limiting function. Of course, the decision outside the **neuralNetwork.vi** comes from the *sumOut* and evaluates this value in a symmetric hard limiting case. □

Neurons communicate between themselves and form a neural network. If we use the mathematical neural model, then we can create an ANN. The basic idea behind ANNs is to simulate the behavior of the human brain in order to define an artificial computation and solve several problems. The concept of an ANN introduces a simple form of biological neurons and their interactions, passing information through the links. That information is essentially transformed in a computational way by mathematical models and algorithms.

Neural networks have the following properties:

1. Able to learn data collection;
2. Able to generalize information;
3. Able to recognize patterns;

4. Filtering signals;
5. Classifying data;
6. Is a massively parallel distributed processor;
7. Predicting and approximating functions;
8. Universal approximators.

Considering their properties and applications, ANNs can be classified as: supervised networks, unsupervised networks, competitive or self-organizing networks, and recurrent networks.

As seen above, ANNs are used to generalize information, but first need to be trained. Training is the process where neural models find the weights of each neuron. There are several methods of training like the backpropagation algorithm used in feed-forward networks. The training procedure is actually derived from the need to minimize errors.

For example, if we are trying to find the weights in a supervised network. Then, we have to have at least some input and output data samples. With this data, by different methods of training, ANNs measure the error between the actual output of the neural network and the desired output. The minimization of error is the target of every training procedure. If it can be found (the minimum error) then the weights that produce this minimization are the optimal weights that enable the trained neural network to be ready for use. Some applications in which ANNs have been used are (general and detailed information found in [1–14]):

Analysis in forest industry. This application was developed by O. Simula, J. Vesanto, P. Vasara and R.R. Helminen in Finland. The core of the problem is to cluster the pulp and paper mills of the world in order to determine how these resources are valued in the market. In other words, executives want to know the competitiveness of their packages coming from the forest industry. This clustering was solved with a Kohonen network system analysis.

Detection of aircraft in synthetic aperture radar (SAR) images. This application involves real-time systems and image recognition in a vision field. The main idea is to detect aircrafts in images known as SAR and in this case they are color aerial photographs. A multi-layer neural network perceptron was used to determine the contrast and correlation parameters in the image, to improve background discrimination and register the RGB bands in the images. This application was developed by A. Filippidis, L.C. Jain and N.M. Martin from Australia. They use a fuzzy reasoning in order to benefit more from the advantages of artificial intelligence techniques. In this case, neural networks were used in order to design the inside of the fuzzy controllers.

Fingerprint classification. In Turkey, U. Halici, A. Erol and G. Ongun developed a fingerprint classification with neural networks. This approach was designed in 1999 and the idea was to recognize fingerprints. This is a typical application using ANNs. Some people use multi-layer neural networks and others, as in this case, use self-organizing maps. *Scheduling communication systems.* In the Institute of Informatics and Telecommunications in Italy, S. Cavalieri and O. Mirabella developed a multi-layer neural network system to optimize a scheduling in real-time communication systems.

Controlling engine generators. In 2004, S. Weifeng and T. Tianhao developed a controller for a marine diesel engine generator [2]. The purpose was to implement a controller that could modify its parameters to encourage the generator with optimal behavior. They used neural networks and a typical PID controller structure for this application.

3.2 Artificial Neural Network Classification

Neural models are used in several problems, but there are typically five main problems in which ANNs are accepted (Table 3.1). In addition to biological neurons, ANNs have different structures depending on the task that they are trying to solve. On one hand, neural models have different structures and then, those can be classified in the two categories below. Figure 3.14 summarizes the classification of the ANN by their structures and training procedures.

Feed-forward networks. These neural models use the input signals that flow only in the direction of the output signals. Single and multi-layer neural networks are typical examples of that structure. Output signals are consequences of the input signals and the weights involved.

Feed-back networks. This structure is similar to the last one but some neurons have loop signals, that is, some of the output signals come back to the same neuron or neurons placed before the actual one. Output signals are the result of the non-transient response of the neurons excited by input signals.

On the other hand, neural models are classified by their learning procedure. There are three fundamental types of models, as described in the following:

1. *Supervised networks.* When we have some data collection that we really know, then we can train a neural network based on this data. Input and output signals are imposed and the weights of the structure can be found.

Table 3.1 Main tasks that ANNs solve

Task	Description
Function approximation	Linear and non-linear functions can be approximated by neural networks. Then, these are used as fitting functions.
Classification	1. *Data classification.* Neural networks assign data to a specific class or subset defined. Useful for finding patterns.
	2. *Signal classification.* Time series data is classified into subsets or classes. Useful for identifying objects.
Unsupervised clustering	Specifies order in data. Creates clusters of data in unknown classes.
Forecasting	Neural networks are used to predict the next values of a time series.
Control systems	Function approximation, classification, unsupervised clustering and forecasting are characteristics that control systems uses. Then, ANNs are used in modeling and analyzing control systems.

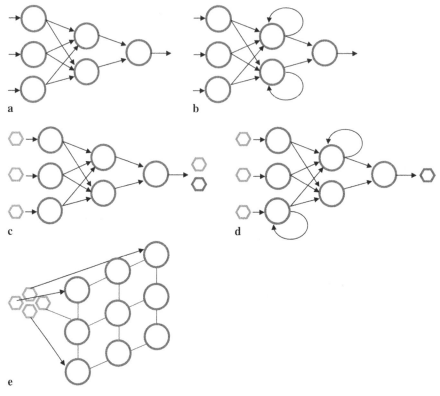

Fig. 3.14a–e Classification of ANNs. **a** Feed-forward network. **b** Feed-back network. **c** Supervised network. **d** Unsupervised network. **e** Competitive or self-organizing network

2. *Unsupervised networks.* In contrast, when we do not have any information, this type of neural model is used to find patterns in the input space in order to train it. An example of this neural model is the Hebbian network.
3. *Competitive or self-organizing networks.* In addition to unsupervised networks, no information is used to train the structure. However, in this case, neurons fight for a dedicated response by specific input data from the input space. Kohonen maps are a typical example.

3.3 Artificial Neural Networks

The human brain adapts its neurons in order to solve the problem presented. In these terms, neural networks shape different architectures or arrays of their neurons. For different problems, there are different structures or models. In this section, we explain the basis of several models such as the perceptron, multi-layer neural networks, trigonometric neural networks, Hebbian networks, Kohonen maps and Bayesian networks. It will be useful to introduce their training methods as well.

3.3.1 Perceptron

Perceptron or threshold neuron is the simplest form of the biological neuron modeling. This kind of neuron has input signals and they are weighted. Then, the activation function decides and the output signal is offered. The main point of this type of neuron is its activation function modeled as a threshold function like that in (3.3). Perceptron is very useful to classify data. As an example, consider the data shown in Table 3.2.

$$f(s) = y = \begin{cases} 0 & s < 0 \\ 1 & s \geq 0 \end{cases} \tag{3.3}$$

We want to classify the input vector $X = \{x_1, x_2\}$ as shown by the target y. This example is very simple and simulates the *AND* operator. Suppose then that weights are $W = \{1, 1\}$ (so-called weight vector) and the activation function is like that given in (3.3). The neural network used is a perceptron. What are the output values for each sample of the input vector at this time?

Create a new VI. In this VI we need a real-value matrix for the input vector X and two 1D arrays. One of these arrays is for the weight vector W and the other is for the output signal **y**. Then, a for-loop is located in order to scan the **X**-matrix row by row. Each row of the **X**-matrix with the weight vector is an inner product implemented with the **sum_weight_inputs.vi** located at ICTL ≫ ANNs ≫ Perceptron ≫ Neuron Parts ≫ **sum_weight_inputs.vi**. The xi connector is for the row vector of the X-matrix, the w_{ij} is for the weight array and the *bias* pin in this moment gets the value 0. The explanation of this parameter is given below. After that, the activation function is evaluated at the sum of the linear combination.

We can find this activation function in the path ICTL ≫ ANNs ≫ Perceptron ≫ Transfer F. ≫ **threshold.vi**. The *threshold* connector is used to define in which value the function is discontinued. Values above this threshold are 1 and values below this one are 0. Finally, these values are stored in the output array. Figure 3.15 shows the block diagram and Fig. 3.16 shows the front panel.

Table 3.2 Data for perceptron example

x_1	x_2	y
0.2	0.2	0
0.2	0.8	0
0.8	0.2	0
0.8	0.8	1

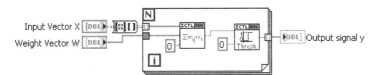

Fig. 3.15 Block diagram for evaluating a perceptron

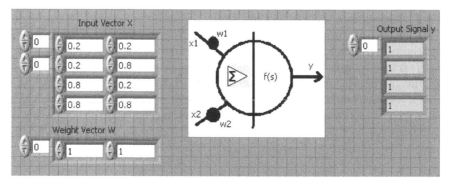

Fig. 3.16 Calculations for the initial state of the perceptron learning procedure

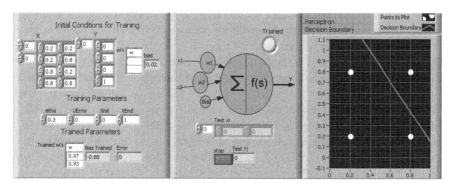

Fig. 3.17 Example of the trained perceptron network emulating the *AND* operator

As we can see, the output signals do not coincide with the values that we want. In the following, the training will be performed as a supervised network. Taking the desired output value y and the actual output signal y', the error function can be determined as in (3.4):

$$E = y - y'. \tag{3.4}$$

The rule of updating the weights is in given as:

$$w_{new} = w_{old} + \eta E X, \tag{3.5}$$

where w_{new} is the updated weight, w_{old} is the actual weight, η is the learning rate, a constant between 0 and 1 that is used to adjust how fast learning is, and $X = \{x_1, x_2\}$ for this example and in general $X = \{x_1, x_2, \ldots, x_n\}$ is the input vector. This rule applies to every single weight participating in the neuron. Continuing with the example for LabVIEW, assume the learning rate is $\eta = 0.3$, then the updating weights are as in Fig. 3.17.

This example can be found in ICTL ≫ ANNs ≫ Perceptron ≫ **Example_Percep tron.vi**. At this moment we know the X-matrix or the 2D array, the desired Y-array. The parameter *etha* is the learning rate, and *UError* is the error that we want to have between the desired output signal and the current output for the perceptron. To draw the plot, the interval is $[Xinit, XEnd]$. The weight array and the *bias* are selected, initializing randomly. Finally, the *Trained Parameters* are the values found by the learning procedure.

In the second block of Fig. 3.17, we find the test panel. In this panel we can evaluate any point $X = \{x_1, x_2\}$ and see how the perceptron classifies it. The Boolean LED is on only when a solution is found. Otherwise, it is off. The third panel in Fig. 3.17 shows the graph for this example. The red line shows how the neural network classifies points. Any point below this line is classified as 0 and all the other values above this line are classified as 1.

About the bias. In the last example, the training of the perceptron has an additional element called *bias*. This is an input coefficient that preserves the action of translating the red line displayed by the weights (it is the cross line that separates the elements). If no bias were found at the neuron, the red line can only move around the zero-point. Bias is used to translate this red line to another place that makes possible the classification of the elements in the input space. As with input signals, bias has its own weight. Arbitrarily, the bias value is considered as one unit. Therefore, bias in the previous example is interpreted as the weight of the unitary value.

This can be viewed in the 2D space. Suppose, $X = \{x_1, x_2\}$ and $W = \{w_1, w_2\}$. Then, the linear combination is done by:

$$y = f\left(\sum_i x_i w_i + b\right) = f(x_1 w_1 + x_2 w_2 + b). \tag{3.6}$$

Then,

$$f(s) = \begin{cases} 0 & \text{if } -b > x_1 w_1 + x_2 w_2 \\ 1 & \text{if } -b \leq x_1 w_1 + x_2 w_2 \end{cases}. \tag{3.7}$$

Then, $\{w_1, w_2\}$ form a basis of the output signal. By this fact, W is orthogonal to the input vector $X = \{x_1, x_2\}$. Finally, if the inner product of these two vectors is zero then we can know that the equations form a boundary line for the decision process. In fact, the boundary line is:

$$x_1 w_1 + x_2 w_2 + b = 0. \tag{3.8}$$

Rearranging the elements, the equation becomes:

$$x_1 w_1 + x_2 w_2 = -b. \tag{3.9}$$

Then, by linear algebra we know that the last equation is the expression of a plane, with distance from the origin equal to $-b$. So, b is in fact the deterministic value that translates the line boundary more closely or further away from the zero-point. The angle for this line between the x-axis is determined by the vector W. In general, the line boundary is plotted by:

$$x_1 w_1 + \ldots + x_n w_n = -b. \tag{3.10}$$

We can make perceptron networks with the condition that neurons have an activation function like that found in (3.3). By increasing the number of perceptron neurons, a better classification of non-linear elements is done. In this case, neurons form

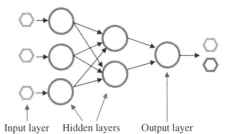

Fig. 3.18 Representation of
a feed-forward multi-layer
neural network

Input layer   Hidden layers   Output layer

layers. Each layer is connected to the next one if the network is feed-forward. In another case, layers can be connected to their preceding or succeeding layers. The first layer in known as the *input layer*, the last one is the *output layer*, where the intermediate layers are called *hidden layers* (Fig. 3.18).

The algorithm for training a feed-forward perceptron neural network is presented in the following:

Algorithm 3.1	Learning procedure of perceptron nets
Step 1	Determine a data collection of the input/output signals (x_i, y_i). Generate random values of the weights w_i. Initialize the time $t = 0$.
Step 2	Evaluate perceptron with the inputs x_i and obtain the output signals y'_i.
Step 3	Calculate the error E with (3.4).
Step 4	If error $E = 0$ for every i then *STOP*. Else, update weight values with (3.5), $t \leftarrow t + 1$ and go to *Step 2*.

3.3.2 Multi-layer Neural Network

This neural model is quite similar to the perceptron network. However, the activation function is not a unit step. In this ANN, neurons have any number of activation functions; the only restriction is that functions must be continuous in the entire domain.

3.3.2.1 ADALINE

The easiest neural network is the adaptive linear neuron (ADALINE). This is the first model that uses a linear activation function like $f(s) = s$. In other words, the inner product of the input and weight vectors is the output signal of the neuron. More precisely, the function is as in (3.11):

$$y = f(s) = s = w_0 + \sum_{i=1}^{n} w_i x_i , \qquad (3.11)$$

where w_0 is the bias weight. Thus, as with the previous networks, this neural network needs to be trained. The training of this neural model is called the delta rule. In this case, we assume one input x to a neuron. Thus, considering an ADALINE, the error is measured as:

$$E = y - y' = y - w_1 x .$$ (3.12)

Looking for the square of the error, we might have

$$e = \frac{1}{2} (y - w_1 x)^2 .$$ (3.13)

Trying to minimize the error is the same as the derivative of the error with respect to the weight, as shown in (3.14):

$$\frac{de}{dw} = -E x .$$ (3.14)

Thus, this derivative tells us in which direction the error increases faster. The weight change must then be proportional and negative to this derivative. Therefore, $\Delta w = \eta E x$, where η is the learning rate. Extending the updating rule of the weights to a multi-input neuron is show in (3.15):

$$w_0^{t+1} = w_0^t + \eta E$$
$$w_i^{t+1} = w_i^t + \eta E x_i .$$ (3.15)

A supervised ADALINE network is used if a threshold is placed at the output signal. This kind of neural network is known as a linear multi-layer neural network without saturation of the activation function.

3.3.2.2 General Neural Network

ADALINE is a linear neural network by its activation function. However, in some cases, this activation function is not the desirable one. Other functions are then used, for example, the sigmoidal or the hyperbolic tangent functions. These functions are shown in Fig. 3.3.

In this way, the delta rule cannot be used to train the neural network. Therefore another algorithm is used based on the gradient of the error, called the backpropagation algorithm. We need a pair of input/output signals to train the neural model. This type of ANN is then classified as supervised and feed-forward, because the input signals go from the beginning to the end.

When we are attempting to find the error between the desired value and the actual value, only the error at the last layer (or the output layer) is measured. Therefore, the idea behind the backpropagation algorithm is to retro-propagate the error from the output layer to the input layer through hidden layers. This ensures that a *kind* of proportional error is preserved in each neuron. The updating of the weights can then be done by a variation or delta error, proportional to a learning rate.

First, we divide the process into two structures. One is for the values at the last layer (output layer) and the other values are from the hidden layers to the input layers. In these terms, the updating rule of the output weights is

$$\Delta v_{ji} = \sum_j \left(-\eta \delta_j^q z_i^q \right),$$
(3.16)

where v_{ji} is the weight linking the ith actual neuron with the jth neuron in the previous layer, and q is the number of the sample data. The other variables are given in (3.17):

$$z_i^q = f \left(\sum_{k=0}^{n} w_{ik} x_k^q \right).$$
(3.17)

This value is the input to the hidden neuron i in (3.18):

$$\delta_j^q = \left(o_j^q - y_j^q \right) f' \left(\sum_{k=1}^{m} v_{jk} z_k^q \right).$$
(3.18)

Computations of the last equations come from the delta rule. We also need to understand that in hidden layers there are no desired values to compare. Then, we propagate the error to the last layers in order to know how neurons produce the final error. These values are computed by:

$$\Delta^q w_{ik} = -\eta \frac{\partial E^q}{\partial w_{ik}} = -\eta \frac{\partial E^q}{\partial o_i^q} \frac{\partial o_i^q}{\partial w_{ik}},$$
(3.19)

where o_i^q is the output of the ith hidden neuron. Then, $o_i^q = z_i^q$ and

$$\frac{\partial o_i^q}{\partial w_{ik}} = f' \left(\sum_{h=0}^{n} w_{ih} x_h^q \right) x_k^q.$$
(3.20)

Now, we obtain the value

$$\delta_i^q = \frac{\partial E^q}{\partial o_i^q} = \sum_{j=1}^{g} \frac{\partial E^q}{\partial o_j^q} \frac{\partial o_j^q}{\partial o_i^q},$$
(3.21)

which is related to the hidden layer. Observe that j is the element of the jth output neuron. Finally, we already know the values $\frac{\partial E^q}{\partial o_j^q}$ and the last expression is:

$$\delta_i^q = f_i' \left(\sum_{k=0}^{n} w_{ik} x_k^q \right) \sum_{j=1}^{p} v_{ij} \delta_j^q.$$
(3.22)

Algorithm 3.2 shows the backpropagation learning procedure for a two-layer neural network (an input layer, one hidden layer, and the output layer). This algorithm can be easily extended to more than one hidden layer. The last net is called a multi-layer or n-layer feed-forward neural network. Backpropagation can be thought of as a generalization of the delta rule and can be used instead when ADALINE is implemented.

Algorithm 3.2	Backpropagation
Step 1	Select a learning rate value η. Determine a data collection of q samples of inputs x and outputs y. Generate random values of weights w_{ik} where i specifies the ith neuron in the actual layer and k is the kth neuron of the previous layer. Initialize the time $t = 0$.
Step 2	Evaluate the neural network and obtain the output values o_i.
Step 3	Calculate the error as $E^q(w) = \frac{1}{2} \sum_{i=1}^{P} (o_i^q - y_i^q)^2$.
Step 4	Calculate the delta values of the output layer: $\delta_i^q = f_i'(\sum_{k=1}^{n} v_{ik} z_k)(o_i^q - y_i^q)$. Calculate the delta values at the hidden layer as: $\delta_i^q = f_i'(\sum_{k=0}^{n} w_{ik} x_k^q) \sum_{j=1}^{P} v_{ij} \delta_j^q$.
Step 5	Determine the change of weights as $\Delta w_{ik}^q = -\eta \delta_i^q o_k^q$ and update the parameters with the next rule $w_{ik}^q \leftarrow w_{ik}^q + \Delta w_{ik}^q$.
Step 6	If $E \leq e$ min where e min is the minimum error expected then *STOP*. Else, $t \leftarrow t + 1$ and go to *Step 2*.

Example 3.2. Consider the points in $\mathfrak{R}^2$ as in Table 3.3. We need to classify them into two clusters by a three-layer feed-forward neural network (with one hidden layer). The last column of the data represents the target $\{0, 1\}$ of each cluster. Consider the learning rate to be 0.1.

Table 3.3 Data points in $\mathfrak{R}^2$

Point	X-coordinate	Y-coordinate	Cluster
1	1	2	0
2	2	3	0
3	1	1	0
4	1	3	0
5	2	2	0
6	6	6	1
7	7	6	1
8	7	5	1
9	8	6	1
10	8	5	1

Solution. First, we have the input layer with two neurons; one for the x-coordinate and the second one for the y-coordinate. The output layer is simply a neuron that must be in the domain $[0, 1]$. For this example we consider a two-neuron hidden layer (actually, there is no analytical way to define the number of hidden neurons).

Table 3.4 Randomly initialized weights

Weights between the first and second layers	Weights between the second and third layers
0.0278	0.0004
0.0148	0.0025
0.0199	
0.0322	

We need to consider the following parameters:

Activation function:	Sigmoidal
Learning rate:	0.1
Number of layers:	3
Number of neurons per layer:	$2 - 2 - 1$

Other parameters that we need to consider are related to the stop criterion:

Maximum number of iterations:	1000
Minimum error or energy:	0.001
Minimum tolerance of error:	0.0001

In fact, the input training data are the two columns of coordinates. The output training data is the last column of cluster targets. The last step before the algorithm will train the net is to initialize the weights randomly. Consider as an example, the randomizing of values in Table 3.4.

According to the above parameters, we are able to run the backpropagation algorithm implemented in LabVIEW. Go to the path ICTL ≫ ANNs ≫ Backpropagation ≫ **Example_Backpropagation.vi**. In the front panel, we can see the window shown in Fig. 3.19. Desired input values must be in the form of (3.23):

$$X = \begin{bmatrix} x_1^1 & \cdots & x_1^m \\ \vdots & \ddots & \vdots \\ x_n^1 & \cdots & x_n^m \end{bmatrix}, \tag{3.23}$$

where $x^j = \{x_1^j, \ldots, x_n^j\}^T$ is the column vector of the jth sample with n elements. In our example, $x^j = \{X^j, Y^j\}$ has two elements. Of course, we have 10 samples of that data, so $j = 1, \ldots, 10$. The desired input data in the matrix looks like Fig. 3.20. The desired output data must also be in the same form as (3.23).

The term $y^j = \{y_1^j, \ldots, y_r^j\}^T$ is the column of the jth sample with r elements. In our example, we have $y^j = \{C^j\}$, where C is the corresponding value of the cluster. In fact, we need exactly $j = 1, \ldots, 10$ terms to solve the problem. This matrix looks like Fig. 3.21.

In the *function* value we will select *Sigmoidal*. In addition, L is the number of layers in the neural network. We treated a three-layer neural network, so $L = 3$. The

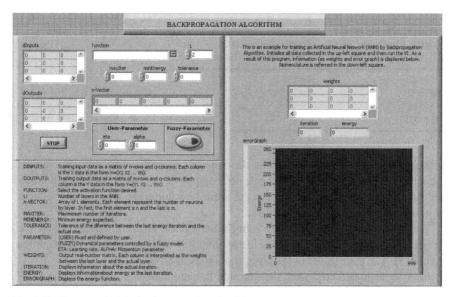

Fig. 3.19 Front panel of the backpropagation algorithm

dInputs										
1	2	1	1	2	6	7	7	8	8	0
2	3	1	3	2	6	6	5	6	5	0
0	0	0	0	0	0	0	0	0	0	0

Fig. 3.20 Desired input data

dOutputs										
0	0	0	0	0	1	1	1	1	1	0
0	0	0	0	0	0	0	0	0	0	0

Fig. 3.21 Desired output data

n-vector is an array in which each of the elements represents the number of neurons per layer. Indeed, we have to write the array *n-vector* $= \{2, 2, 1\}$. Finally, *maxIter* is the maximum number of iterations we want to wait until the best answer is found. *minEnergy* is the minimum error between the desired output and the actual values derived from the neural network.

Tolerance is the variable that controls the minimum change in error that we want in the training procedure. Then, if one of the three last values is reached, the procedure will stop. We can use crisp parameters of fuzzy parameters to train the network, where *eta* is the learning rate and *alpha* is the momentum parameter.

As seen in Fig. 3.19, the right window displays the result. *Weights* values will appear until the process is finished and there are the coefficients of the trained neural

Table 3.5 Trained weights

Weights between the first and second layers	Weights between the second and third layers
0.3822	1.8230
−0.1860	1.8710
0.3840	
−0.1882	

network. The *errorGraph* shows the decrease in the error value when the actual output values are compared with the desired output values. The real-valued number appears in the *error* indicator. Finally, the *iteration* value corresponds to the number of iterations completed at the moment.

With those details, the algorithm is implemented and the training network (or the weights) is shown in Table 3.5 (done in 184 iterations and reaching the local minima at 0.1719). The front panel of the algorithm looks like Fig. 3.22.

In order to understand what this training has implemented, there are graphs of this classification. In Fig. 3.23, the first graph is the data collection, and the second graph shows the clusters. If we see a part of the block diagram in Fig. 3.24, only the input data is used in the three-layer neural network. To show that this neural network can generalize, other data different from the training collection is used. Looking at Fig. 3.25, we see the data close to the training zero-cluster.   □

When the learning rate is not selected correctly, the solution might be trapped in local minima. In other words, minimization of the error is not reached. This can be

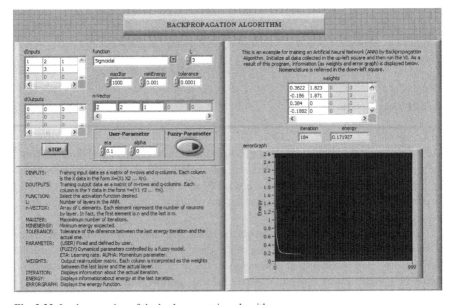

Fig. 3.22 Implementation of the backpropagation algorithm

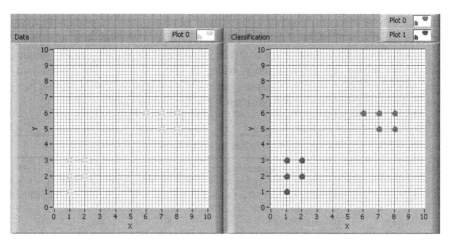

Fig. 3.23 The *left* side shows a data collection, and the *right* shows the classification of that data

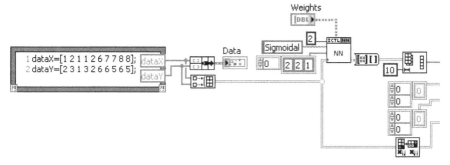

Fig. 3.24 Partial view of the block diagram in classification data, showing the use of the neural network

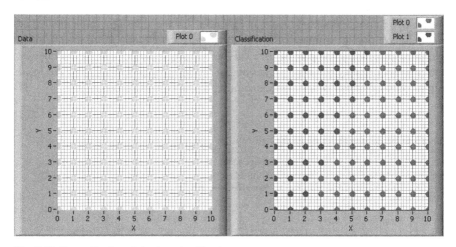

Fig. 3.25 Generalization of the data classification

partially solved if the learning rate is decreased, but time grows considerably. One solution is the modification of the backpropagation algorithm by adding a momentum coefficient. This is used to try to get the tending of the solution in the weight space. This means that the solution is trying to find and follow the tendency of the previous updating weights. That modification is summarized in Algorithm 3.3, which is a rephrased version of Algorithm 3.2 with the new value.

Algorithm 3.3	Backpropagation with momentum parameter
Step 1	Select a learning rate value η and momentum parameter α. Determine a data collection of q samples of inputs x and outputs y. Generate random values of weights w_{ik} where i specifies the ith neuron in the actual layer and k is the kth neuron of the previous layer. Initialize the time $t = 0$.
Step 2	Evaluate the neural network and obtain the output values o_i.
Step 3	Calculate the error as $E^q(w) = \frac{1}{2} \sum_{i=1}^{P} (o_i^q - y_i^q)^2$.
Step 4	Calculate the delta values of the output layer: $\delta_i^q = f_i'(\sum_{k=1}^{n} v_{ik} z_k)(o_i^q - y_i^q)$. Calculate the delta values at the hidden layer as: $\delta_i^q = f_i'(\sum_{k=0}^{n} w_{ik} x_k^q) \sum_{j=1}^{P} v_{ij} \delta_j^q$.
Step 5	Determine the change of weights as $\Delta w_{ik}^q = -\eta \delta_i^q o_k^q$ and update the parameters with the next rule: $w_{ik}^q \leftarrow w_{ik}^q + \Delta w_{ik}^q + \alpha \left(w_{ik\_act}^q - w_{ik\_last}^q \right)$ where w_{act} is the actual weight and w_{last} is the previous weight.
Step 6	If $E \leq e$ min where e min is the minimum error expected then *STOP*. Else, $t \leftarrow t + 1$ and go to *Step 2*.

Example 3.3. Train a three-layer feed-forward neural network using a 0.7 momentum parameter value and all data used in Example 3.2.

Solution. We present the final results in Table 3.6 and the algorithm implemented in Fig. 3.26. We find the number of iterations to be 123 and the local minima 0.1602, with a momentum parameter of 0.7. This minimizes in some way the number of iterations (decreasing the time processing at the learning procedure) and the local minima is smaller than when no momentum parameter is used.   □

Table 3.6 Trained weights for feed-forward network

Weights between the first and second layers	Weights between the second and third layers
0.3822	1.8230
−0.1860	1.8710
0.3840	
−0.1882	

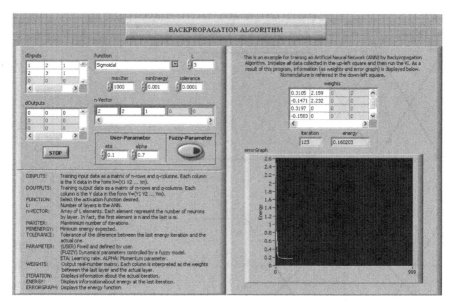

Fig. 3.26 Implementation of the backpropagation algorithm with momentum parameter

3.3.2.3 Fuzzy Parameters in the Backpropagation Algorithm

In this section we combine the knowledge about fuzzy logic and ANNs. In this way, the main idea is to control the parameters of learning rate and momentum in order to get fuzzy values and then evaluate the optimal values for these parameters.

We first provide the fuzzy controllers for the two parameters at the same time. As we know from Chap. 2 on fuzzy logic, we evaluate the error and the change in the error coefficients from the backpropagation algorithm. That is, after evaluating the error in the algorithm, this value enters the fuzzy controller . The change in the error is the difference between the actual error value and the last error evaluated.

Input membership functions are represented as the normalized domain drawn in Figs. 3.27 and 3.28. Fuzzy sets are *low positive* (LP), *medium positive* (MP), and *high positive* (HP) for error value E. In contrast, fuzzy sets for change in error CE are *low negative* (LN), *medium negative* (MN), and *high negative* (HN). Figure 3.29 reports the fuzzy membership functions of change parameter $\Delta\beta$ with fuzzy sets *low negative* (LN), *zero* (ZE), and *low positive* (LP). Tables 3.7 and 3.8 have the fuzzy associated matrices (FAM) to imply the fuzzy rules for the learning rate and momentum parameter, respectively.

In order to access the fuzzy parameters, go to the path ICTL ≫ ANNs ≫ Backpropagation ≫ **Example_Backpropagation.vi**. As with previous examples, we can obtain better results with these fuzzy parameters. Configure the settings of this VI except for the learning rate and momentum parameter. Switch on the *Fuzzy-Parameter* button and run the VI. Figure 3.30 shows the window running this configuration.

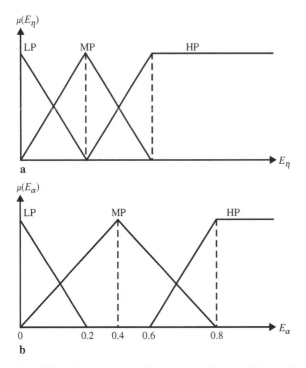

Fig. 3.27a,b Input membership functions of error. **a** Error in learning parameter. **b** Error in momentum parameter

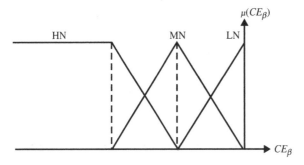

Fig. 3.28 Input membership functions of change in error

Table 3.7 Rules for changing the learning rate

$E \backslash CE$	LN	MN	HN
LP	ZE	ZE	LN
MP	LP	ZE	ZE
HP	LP	LP	ZE

Fig. 3.29 Output membership functions of change in the parameter selected

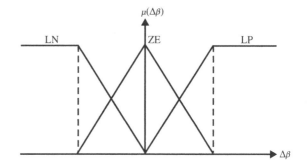

Table 3.8 Rules for changing the momentum parameter

$E \setminus CE$	LN	MN	HN
LP	ZE	LN	LN
MP	ZE	LN	LN
HP	LP	ZE	ZE

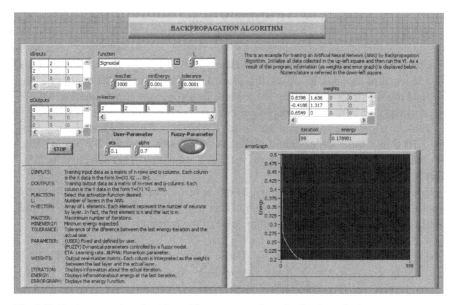

Fig. 3.30 Backpropagation algorithm with parameter adjusted using fuzzy logic

3.3.3 Trigonometric Neural Networks

In the previous neural networks, we saw that supervised and feed-forward neural models need to be trained by iterative methods. This situation increases the time of convergence of the learning procedure. In this section, we introduce a trigonometric-based neural network.

First, as we know, a Fourier series is used to approximate a periodic function $f(t)$ with constant period T. It is well known that any function can be approximated by a Fourier series, and so this type of network is used for periodic signals.

Consider a function as in (3.24):

$$f(t) = \frac{1}{2}a_0 + a_1 \cos \omega_0 t + a_2 \cos 2\omega_0 t + \ldots + b_1 sen \omega_0 t + b_2 sen 2\omega_0 t + \ldots$$

$$f(t) = \frac{1}{2}a_0 + \sum_{n=1}^{\infty} [a_n \cos(n\omega_0 t) + b_n sen(n\omega_0 t)]$$

$$f(t) = C_0 + \sum_{n=1}^{\infty} C_n \cos(n\omega_0 t - \Theta n). \tag{3.24}$$

Looking at the neural networks described above, this series is very similar to the mathematical neural model when the activation function is linear:

$$y = x_0 + \sum_{i=1}^{n} w_i x_i. \tag{3.25}$$

Comparing (3.24) and (3.25), we see that they are very close in form, except for the infinite terms of the sum. However, this is not a disadvantage. On the contrary, if we truncate the sum to N terms, then we produce an error in the approximation. This is clearly helpful in neural networks because we do not need them to be memorized.

Thus, a trigonometric neural network (T-ANN) is a Fourier-based net . Figure 3.30 shows this type of neural model. As we might suppose, T-ANN are able to compute with cosine functions or with sine functions. This selection is arbitrary.

Considering its learning procedure, a Fourier series can be solved analytically by employing least square estimates (LSE). This process means that we want to find coefficients that preserve the minimum value of the function

$$S(a_0, a_1, \ldots, a_n) = \sum_{i=1}^{m} \left[y_i - \left(\frac{1}{2}a_0 + \sum_{k=1}^{\infty} a_k \cos(k\omega_0 x_i) \right) \right]^2, \tag{3.26}$$

where ω_0 is the fundamental frequency of the series, x_i is the ith input data and y_i is the ith value of the desired output. Then, we need the first derivative of that function, which is:

$$\frac{\delta S}{\delta a_p} = \sum_{i=1}^{m} \left[y_i - \left(\frac{1}{2}a_0 + \sum_{k=1}^{\infty} a_k \cos(k\omega_0 x_i) \right) \cos(n\omega_0 x) \right] = 0, \quad \forall p \geq 1. \tag{3.27}$$

This is a system of linear equations that can be viewed as:

$$
\begin{bmatrix}
\frac{1}{2}m & \cdots & \sum_{i=1}^{m}\cos(p\omega_0 x_i) \\
\frac{1}{2}\sum_{i=1}^{m}\cos(\omega_0 x_i) & \cdots & \sum_{i=1}^{m}\cos(\omega_0 x_i)\cos(p\omega_0 x_i) \\
\vdots & \ddots & \sum_{i=1}^{m}\cos(p\omega_0 x_i)\cos(p\omega x_i) \\
\frac{1}{2}\sum_{i=1}^{m}\cos(p\omega_0 x_i) & \cdots & \sum_{i=1}^{m}\cos^2(p\omega_0 x_i)
\end{bmatrix}
\begin{bmatrix}
a_0 \\ a_1 \\ \vdots \\ a_n
\end{bmatrix}
$$

$$
=
\begin{bmatrix}
\sum_{i=1}^{m} y_i \\
\sum_{i=1}^{m} y_i \cos(\omega_0 x_i) \\
\vdots \\
\sum_{i=1}^{m} y_i \cos(p\omega_0 x_i)
\end{bmatrix}.
\tag{3.28}
$$

Then, we can solve this system for all coefficients. At this point, p is the number of neurons that we want to use in the T-ANN . In this way, if we have a data collection of the input/output desired values, then we can compute analytically the coefficients of the series or what is the same, the weights of the net. Algorithm 3.4 is proposed for training T-ANNs ; eventually, this procedure can be computed with the backpropagation algorithm as well.

Algorithm 3.4	T-ANNs
Step 1	Determine input/output desired samples. Specify the number of neurons N.
Step 2	Evaluate weights C_i by LSE.
Step 3	*STOP.*

Example 3.4. Approximate the function $f(x) = x^2 + 3$ in the interval $[0, 5]$ with: (a) 5 neurons, (b) 10 neurons, (c) 25 neurons. Compare them with the real function. *Solution.* We need to train a T-ANN and then evaluate this function in the interval $[0, 5]$. First, we access the VI that trains a T-ANN following the path ICTL ≫ ANNs ≫ T-ANN ≫ **entrenaRed.vi**. This VI needs the **x**-vector coordinate, **y**-vector coordinate and the number of neurons that the network will have.

In these terms, we have to create an array of elements between $[0, 5]$ and we do this with a stepsize of 0.1, by the **rampVector.vi**. This array evaluates the function $x^2 + 3$ with the program inside the for-loop in Fig. 3.31. Then, the array coming from the **rampVector.vi** is connected to the x pin of the **entrenaRed.vi**, and the array coming from the evaluated **x**-vector is connected to the y pin. Actually, the pin n is available for the number of neurons. Then, we create a control variable for neurons because we need to train the network with a different number of neurons.

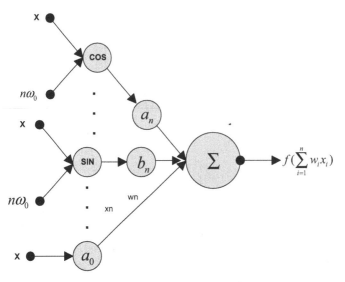

Fig. 3.31 T-ANN model

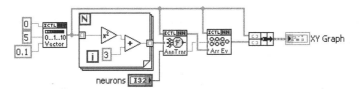

Fig. 3.32 Block diagram of the training and evaluating T-ANN

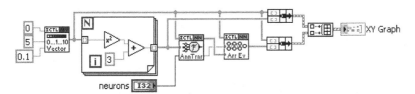

Fig. 3.33 Block diagram for plotting the evaluating T-ANN against the real function

This VI is then connected to another VI that returns the values of a T-ANN. This last node is found in the path ICTL ≫ ANNs ≫ T-ANN ≫ **Arr_Eval_T-ANN.vi**. This receives the coefficients that were the result of the previous VI named *T-ANN Coeff* pin connector. The *Fund Freq* connector is referred to the fundamental frequency of the trigonometric series ω_0. This value is calculated in the **entrenaRed.vi**. The last pin connector is referred to as *Values*. This pin is a 1D array with the values in the *x*-coordinate, which we want to evaluate the neural network. The result of this VI is the output signal of the T-ANN by the pin *T-ANN Eval*. The block diagram of this procedure is given in Fig. 3.32.

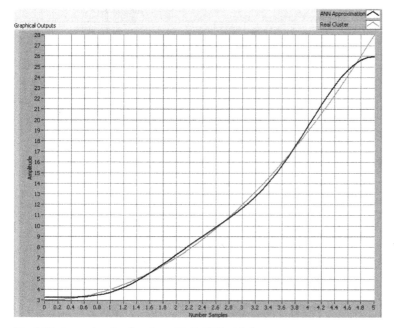

Fig. 3.34 Approximation function with T-ANN with 5 neurons

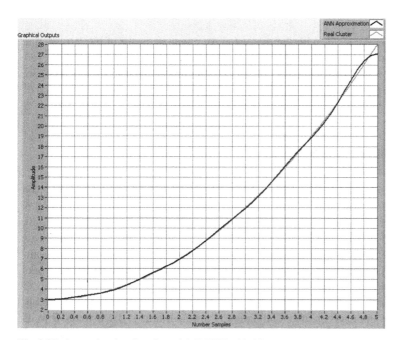

Fig. 3.35 Approximation function with T-ANN with 10 neurons

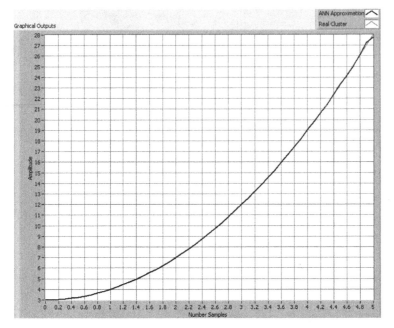

Fig. 3.36 Approximation function with T-ANN with 25 neurons

To compare the result with the real value we create a cluster of two arrays, one comes from the **rampVector.vi** and the other comes from the output of the for-loop. Figure 3.33 shows the complete block diagram. As seen in Figs. 3.34–3.36, the larger the number of neurons, the better the approximation. To generate each of these graphs, we only vary the value of neurons.   □

3.3.3.1 Hebbian Neural Networks

A Hebbian neural network is an unsupervised and competitive net. As unsupervised networks, these only have information about the input space, and their training is based on the fact that the weights store the information. Thus, the weights can only be reinforced if the input stimulus provides sufficient output values. In this way, weights only change proportionally to the output signals. By this fact, neurons compete to become a dedicated reaction of part of the input. Hebbian neural networks are then considered as the first self-organizing nets .

The learning procedure is based on the following statement pronounced by Hebb: As A becomes more efficient at stimulating B during training, A sensitizes B to its stimulus, and the weight on the connection from A to B increases during training as B becomes sensitized to A.

Steven Grossberg then developed a mathematical model for this sentence, given in (3.29):

$$w_{AB}^{\text{new}} = w_{AB}^{\text{old}} + \beta x_B x_A \,, \tag{3.29}$$

where w_{AB} is the weight between the interaction of two neurons A and B, x_i is the output signal of the ith neuron, and $x_B x_A$ is the so-called Hebbian learning term. Algorithm 3.5 introduces the Hebbian learning procedure.

Algorithm 3.5	Hebbian learning procedure
Step 1	Determine the input space.
	Specify the number of iterations $iterNum$ and initialize $t = 0$.
	Generate small random values of weights w_i.
Step 2	Evaluate the Hebbian neural network and obtain the outputs x_i.
Step 3	Apply the updating rule (3.29).
Step 4	If $t = iterNum$ then *STOP*.
	Else, go to *Step 2*.

These types of neural models are good when no desired output values are known. Hebbian learning can be applied in multi-layer structures as well as feed-forward and feed-back networks.

Example 3.5. There are points in the following data. Suppose that this data is some input space. Apply Algorithm 3.5 with a forgotten factor of 0.1 to train a Hebbian network that approximates the data presented in Table 3.9 and Fig. 3.37.

Table 3.9 Data points for the Hebbian example

X-coordinate	Y-coordinate
0	1
1	0
2	2
3	0
4	3.4
5	0.2

Solution. We consider a 0.1 of the learning rate value. The forgotten factor α is applied with the following equation:

$$w_{AB}^{\text{new}} = w_{AB}^{\text{old}} - \alpha w_{AB}^{\text{old}} + \beta x_B x_A \,. \tag{3.30}$$

We go to the path ICTL $\gg$ ANNs $\gg$ Hebbian $\gg$ **Hebbian.vi**. This VI has input connectors of the y-coordinate array, called x pin, which is the array of the desired values, the forgotten factor a, the learning rate value b, and the *Iterations* variable.

Fig. 3.37 Input training data

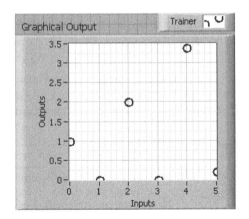

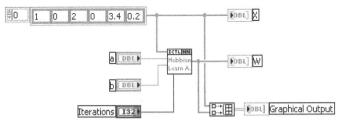

Fig. 3.38 Block diagram for training a Hebbian network

This last value is selected in order to perform the training procedure by this number of cycles. The output of this VI is the weight vector, which is the y-coordinate of the approximation to the desired values. The block diagram for this procedure is shown in Fig. 3.38.

Then, using Algorithm 3.5 with the above rule with forgotten factor, the result looks like Fig. 3.39 after 50 iterations. The vector W is the y-coordinate approximation of the y-coordinate of the input data. Figure 3.39 shows the training procedure.   □

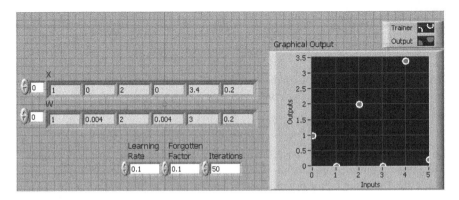

Fig. 3.39 Result of the Hebbian process in a neural network

3.3.4 Kohonen Maps

Kohonen networks or self-organizing maps are a competitive training neural network aimed at ordering the mapping of the input space. In competitive learning, we normally have distributed input $x = x(t) \in \Re^n$, where t is the time coordinate, and a set of reference vectors $m_i = m_i(t) \in \Re^n, \forall i = 1, \ldots, k$. The latter are initialized randomly. After that, given a metric $d(x, m_i)$ we try to minimize this function to find a reference vector that best matches the input. The best reference vector is named m_c (the winner) where c is the best selection index. Thus, $d(x, m_c)$ will be the minimum metric. Moreover, if the input x has a density function $p(x)$, then, we can minimize the error value between the input space and the set of reference vectors, so that all m_i can represent the form of the input as much as possible. However, only an iterative process should be used to find the set of reference vectors.

At each iteration, vectors are actualized by the following equation:

$$m_i(t+1) = \begin{cases} m_i(t) + \alpha(t) \cdot d\,[x(t), m_i(t)] & i = c \\ m_i(t) & i \neq c \end{cases}, \qquad (3.31)$$

where $\alpha(t)$ is a monotonically decreasing function with scalar values between 0 and 1. This method is known as vector quantization (VQ) and looks to minimize the error, considering the metric as a Euclidean distance with r-power:

$$E = \int \|x - m_c\|^r \, p(x)\,dx\,. \qquad (3.32)$$

On the other hand, years of studies on the cerebral cortex have discovered two important things: (1) the existence of specialized regions, and (2) the ordering of these regions. Kohonen networks create a competitive algorithm based on these facts in order to adjust specialized neurons into subregions of the input space, and if this input is ordered, specialized neurons also perform an ordering space (mapping). A typical Kohonen network N is shown in Fig. 3.40.

If we suppose an n-dimensional input space X is divided into subregions x_i, and a set of neurons with a d-dimensional topology, where each neuron is associated to a n-dimensional weight m_i (Fig. 3.40), then this set of neurons forms a space N. Each subregion of the input will be mapped by a subregion of the neuron space. Moreover, mapped subregions will have a specific order because input subregions have order as well.

Kohonen networks emulate the behavior described above, which is defined in Algorithm 3.6.

As seen in the previous algorithm, VQ is used as a basis. To achieve the goal of ordering the weight vectors, one might select the winner vector and its neighbors to approximate the interesting subregion. The number of neighbors v should be a monotonically decreasing function with the characteristic that at the first iteration the network will order uniformly, and then, just the winner neuron will be reshaped to minimize the error.

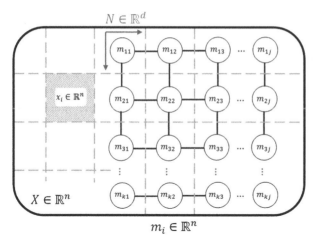

$$m_i \in \mathbb{R}^n$$

Fig. 3.40 Kohonen network N approximating the input space X

Algorithm 3.6	Kohonen learning procedure
Step 1	Initialize the number of neurons and the dimension of the Kohonen network.
	Associate a weight vector m_i to each neuron, randomly.
Step 2	Determine the configuration of the neighborhood N_c of the weight vector considering the number of neighbors v and the neighborhood distribution $v(c)$.
Step 3	Randomly, select a subregion of the input space $x(t)$ and calculate the Euclidean distance to each weight vector.
Step 4	Determine the winner weight vector m_c (the minimum distance defines the winner) and actualize each of the vectors by (3.31) which is a discrete-time notation.
Step 5	Decrease the number of neighbors v and the learning parameter α.
Step 6	Use a statistical parameter to determine the approximation between neurons and the input space. If neurons approximate the input space then *STOP*.
	Else, go to *Step 2*.

Moreover, the training function or learning parameter will be decreased. Figure 3.41 shows how the algorithm is implemented. Some applications of this kind of network are: pattern recognition, robotics, control process, audio recognition, telecommunications, etc.

Example 3.6. Suppose that we have a square region in the interval $x \in [-10, 10]$ and $y \in [-10, 10]$. Train a 2D-Kohonen network in order to find a good approximation to the input space.

Solution. This is an example inside the toolkit, located in ICTL ≫ ANNs ≫ Kohonen SOM ≫ **2DKohonen_Example.vi**. The front panel is the same as in Fig. 3.42, with the following sections.

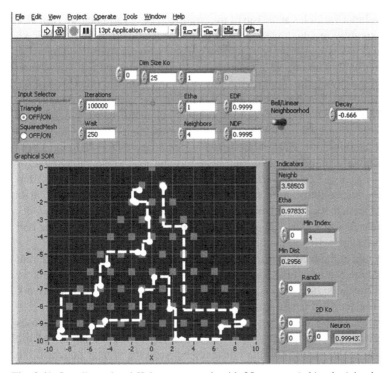

Fig. 3.41 One-dimensional Kohonen network with 25 neurons (*white dots*) implemented to approximate the triangular input space (*red subregions*)

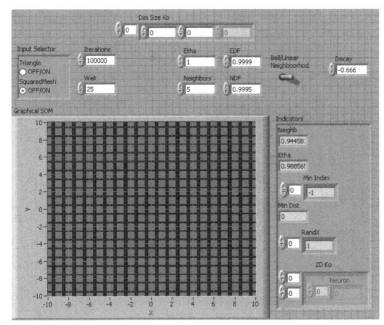

Fig. 3.42 Front panel of the 2D-Kohonen example

We find the input variables at the top of the window. These variables are *Dim Size Ko*, which is an array in which we represent the number of neurons per coordinate system. In fact, this is an example of a 2D-Kohonen network, and the dimension of the Kohonen is 2. This means that it has an *x*-coordinate and a *y*-coordinate. In this case, if we divide the input region into 400 subregions, in other words, we have an interval of 20 elements per 20 elements in a square space, then we may say that we need 20 elements in the *x*-coordinate and 20 elements in the *y*-coordinate dimension. Thus, we are asking for the network to have 400 nodes.

Etha is the learning rate, *EDF* is the learning rate decay factor, *Neighbors* represents the number of neighbors that each node has and its corresponding *NDF* or neighbor decay factor. *EDF* and *NDF* are scalars that decrease the value of *Etha* and *Neighbors*, respectively, at each iteration. After that we have the *Bell/Linear Neighborhood* switch. This switches the type of neighborhood between a bell function and a linear function. The value *Decay* is used as a factor of fitness in the bell function. This has no action in the linear function.

On the left side of the window is the *Input Selector*, which can select two different input regions. One is a triangular space and the other is the square space treated in this example. The value *Iterations* is the number of cycles that the Kohonen network takes to train the net. *Wait* is just a timer to visualize the updating network.

Finally, on the right side of the window is the *Indicators* cluster. It rephrases values of the actual *Neighbor* and *Etha*. *Min Index* represents the indices of the winner node. *Min Dist* is the minimum distance between the winner node and the

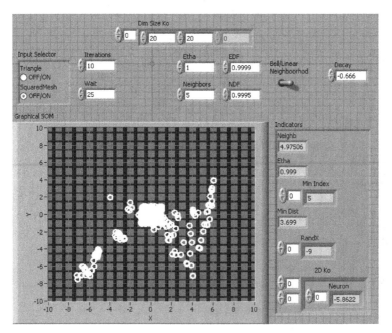

Fig. 3.43 The 2D-Kohonen network at 10 iterations

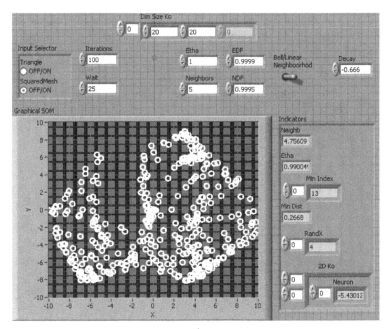

Fig. 3.44 The 2D-Kohonen network at 100 iterations

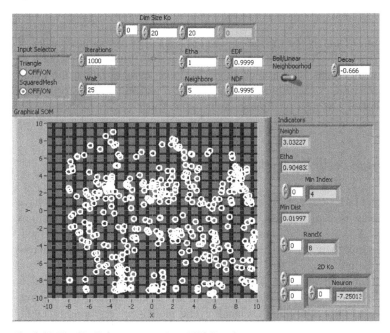

Fig. 3.45 The 2D-Kohonen network at 1000 iterations

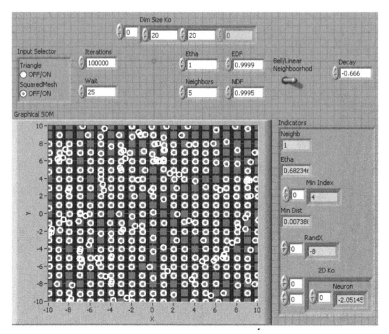

Fig. 3.46 The 2D-Kohonen network at 10 000 iterations

close subregion. *RandX* is the subregion selected randomly. *2D Ko* is a cluster of
nodes with coordinates. Figures 3.42–3.46 represent the current configuration of
the 2D-Kohonen network with five neighbors and one learning rate at the initial
conditions, with values of 0.9999 and 0.9995 for *EDF* and *NDF*, respectively. The
training was done by a linear function of the neighborhood.   □

3.3.5 Bayesian or Belief Networks

This kind of neural model is a directed acyclic graph (DAG) in which nodes have
random variables. Basically, a DAG consists of nodes and deterministic directions
between links. A DAG can be interpreted as an adjacency matrix in which 0 elements mean no links between two nodes, and 1 means a linking between the i th row
and the j th column.

This model can be divided into polytrees and cyclic graphs. Polytrees are models
in which the evidence nodes or the input nodes are at the top, and the children
are below the structure. On the other hand, cyclic models are any kind of DAG,
when going from one node to another node that has at least another path connecting
these points. Figure 3.47 shows examples of these structures. For instance, we only
consider polytrees in this chapter.

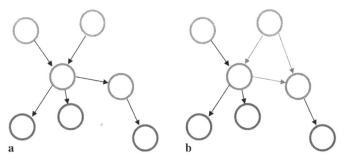

Fig. 3.47a,b Bayesian or belief networks. **a** A polytree. **b** A cyclic structure

Bayesian networks or belief networks have a node V_i that is conditionally independent from a subset of nodes that are not descendents of V_i given its parents $P(V_i)$. Suppose that we have $V_1, \ldots, V_k$ nodes of a Bayesian network and they are conditionally independent. The joint probability of all nodes is:

$$p(V_1, \ldots, V_k) = \prod_{i=1}^{k} p(V_i | P(V_i)). \tag{3.33}$$

These networks are based on tables of probabilities known as conditional probability tables (CPT), in which the node is related to its parents by probabilities.

Bayesian networks can be trained by some algorithms, such as the expectation-maximization (EM) algorithm or the gradient-ascent algorithm. In order to understand the basic idea of training a Bayesian network, a gradient-ascent algorithm will be described in the following.

We are looking to maximize the likelihood hypothesis $\ln P(D|h)$ in which P is the probability of the data D given hypothesis h. This maximization will be performed with respect to the parameters that define the CPT. Then, the expression derived from this fact is:

$$\frac{\partial \ln P(D|h)}{\partial w_{ij}} = \sum_{d \in D} \frac{P(Y_i = y_{ij}, U_i = u_{ik} | d)}{w_{ijk}} \tag{3.34}$$

where y_{ij} is the j-value of the node Y_i, U_i is the parent with the k-value u_{ik}, w_{ijk} is the value of the probability in the CPT relating y_{ij} with u_{ik}, and d is a sample of the training data D. In Algorithm 3.7 this training is described.

Example 3.7. Figure 3.48 shows a DAG. Represent this graph in an adjacency matrix (it is a cyclic structure).
Solution. Here, we present the matrix in Fig. 3.49. Graph theory affirms that the adjacency matrix is unique. Therefore, the solution is unique.   □

Example 3.8. Train the network in Fig. 3.48 for the data sample shown in Table 3.10. Each column represents a node. Note that each node has targets $Y_i = \{0, 1\}$.

Algorithm 3.7	Gradient-ascent learning procedure for Bayesian networks
Step 1	Generate a CPT with random values of probabilities. Determine the learning rate η.
Step 2	Take a sample d of the training data D and determine the probability on the right-hand side of (3.34).
Step 3	Update the parameters with $$w_{ijk} \leftarrow w_{ijk} + \eta \sum_{d \in D} \frac{P(Y_i = y_{ij}, U_i = u_{ik} \mid d)}{w_{ijk}}.$$
Step 4	If $CPT_t = CPT_{t-1}$ then $STOP$. Else, go to $Step\ 2$ until reached.

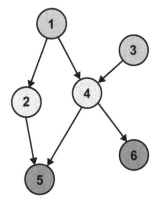

Fig. 3.48 DAG with evidence nodes 1 and 3, and query nodes 5 and 6. The others are known as hidden nodes

Fig. 3.49 Adjacency matrix for the DAG in Fig. 3.48

Table 3.10 Bayesian networks example

Node 1	Node 2	Node 3	Node 4	Node 5	Node 6	Frequency
0	1	1	0	1	1	32
0	1	0	1	0	0	94
0	0	1	0	1	1	83
1	1	0	0	1	0	19
0	0	1	1	0	1	22
1	0	0	0	0	1	18
0	1	1	1	1	0	29
0	0	0	0	1	1	12

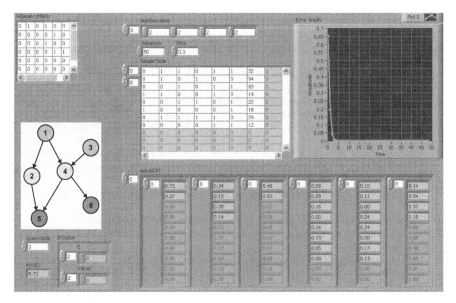

Fig. 3.50 Training procedure of a Bayesian network

Solution. This example is located at ICTL ≫ ANNs ≫ Bayesian ≫ **Bayes_Example.** **vi**. Figure 3.50 shows the implementation of Algorithm 3.7. At the top-left side of the window, we have the adjacency matrix in which we represent the DAG as seen in Example 3.7. Then, *NumberLabels* represents all possible labels that the related node can have. In this case, we have that all nodes can only take values between 0 or 1, then each node has two labels. Therefore, the array is $NumberLabels = \{2, 2, 2, 2, 2, 2\}$. *Iterations* is the same as in the other examples. *Etha* is the learning rate in the gradient-ascent algorithm. *SampleTable* comes from experiments and measures the frequency that some combination of nodes is fired. In this example, the table is the sample data given in the problem.

The *Error Graph* shows how the measure of error is decreasing when time is large. Finally, *ActualCPT* is the result of the training procedure and it is the CPT of the Bayesian network. For instance, we choose a value of learning rate that equals 0.3 and 50 iterations to this training procedure. As we can see, the training needs around five iterations to obtain the CPT. This table contains the training probabilities that relate each node with its immediate parents.   □

References

1. Lakhmi J, Rao V (1999) Industrial applications of neural networks. CRC, Boca Raton, FL
2. Weifeng S, Tianhao T (2004) CMAC Neural networks based combining control for marine diesel engine generator. IEEE Proceedings of the 5th World Congress on Intelligent Control, Hangzshou, China, 15–19 June 2004

3. Ananda M, Srinivas J (2003) Neural networks. Algorithms and applications. Alpha Sciences International, Oxford
4. Samarasinghe S (2007) Neural networks for applied sciences and engineering. Auerbach, Boca Raton, FL
5. Irwin G, et al. (1995) Neural network applications in control. The Institution of Electrical Engineers, London
6. Mitchell T (1997) Machine learning. McGraw-Hill, Boston
7. Kohonen T (1990) The self-organizing map. Proceedings of the IEEE 78(9):1464–1480
8. Rojas R (1996) Neural networks. Kohonen networks, Chap 15. Springer, Berlin Heidelberg New York, pp 391–412
9. Veksler O (2004) Lecture 18: Pattern recognition. University of Western Ontario, Computer Science Department. http://courses.media.mit.edu/2004fall/mas622j. Accessed on 10 March 2009
10. Jensen F (2001) Bayesian networks and decision graphs. Springer, Berlin Heidelberg New York
11. Nilsson N (2001) Inteligencia artificial, una nueva síntesis. McGraw-Hill, Boston
12. Nolte J (2002) The human brain: an introduction to its functional anatomy. Mosby, St. Louis, MO
13. Affi A, Bergman R (2005) Functional neuroanatomy text and atlas. McGraw-Hill, Boston
14. Nguyen H, et al. (2003) A first course in fuzzy and neural control. Chapman & Hall/CRC, London
15. Ponce P (2004) Trigonometric Neural Networks internal report. ITESM-CCM, México City

Futher Reading

Hertz J, Krogh A, Lautrup B, Lehmann T (1997) Nonlinear backpropagation: doing backpropagation without derivatives of the activation function. Neural Networks, IEEE Transactions 8:1321–1327
Loh AP, Fong KF (1993) Backpropagation using generalized least squares. Neural Networks, IEEE International Conference 1:592–597
McLauchlan LLL, Challoo R, Omar SI, McLauchlan RA (1994) Supervised and unsupervised learning applied to robotic manipulator control. American Control Conference, 3:3357–3358
Taji K, Miyake T, Tamura H (1999) On error backpropagation algorithm using absolute error function Systems, Man, and Cybernetics. 1999. IEEE SMC '99 Conference Proceedings, IEEE International Conference 5:401–406

Chapter 4
Neuro-fuzzy Controller Theory and Application

4.1 Introduction

Fuzzy systems allow us to transfer the vague fuzzy form of human reasoning to mathematical systems. The use of *IF–THEN* rules in fuzzy systems gives us the possibility of easily understanding the information modeled by the system. In most of the fuzzy systems the knowledge is obtained from human experts. However this method of information acquisition has a great disadvantage given that not every human expert can and/or want to share their knowledge.

Artificial neural networks (ANNs) can learn from experience but most of the topologies do not allow us to clearly understand the information learned by the networks. ANNs are incorporated into fuzzy systems to form neuro-fuzzy systems, which can acquire knowledge automatically by learning algorithms of neural networks. Neuro-fuzzy systems have the advantage over fuzzy systems that the acquired knowledge, which is easy to understand, is more meaningful to humans.

Another technique used with neuro-fuzzy systems is clustering, which is usually employed to initialize unknown parameters such as the number of fuzzy rules or the number of membership functions for the premise part of the rules. They are also used to create dynamic systems and update the parameters of the system.

An example of neuro-fuzzy systems is the intelligent electric wheelchair. People confined to wheelchairs may get frustrated when attempting to become more active in their communities and societies. Even though laws and pressure from several sources have been made to make cities more accessible to people with disabilities there are still many obstacles to overcome. At Tecnológico de Monterrey Campus Ciudad de México an intelligent electric wheelchair with an autonomous navigation system based on a neuro-fuzzy controller was developed [1, 10].

The basic problem here was that most of the wheelchairs on the market were rigid and failed to adapt to their users, and instead the users had to adapt to the possibilities that the chair gave them. Thus the objective of this project was to create a wheelchair that increased the capabilities of the users, and adapted to every one of them.

P. Ponce-Cruz, F. D. Ramirez-Figueroa, *Intelligent Control Systems with LabVIEW™*   89
© Springer 2010

4.2 The Neuro-fuzzy Controller

Using a neuro-fuzzy controller , the position of the chair is manipulated so that it will avoid static and dynamic obstacles. The controller takes information from three ultrasonic sensors located in different positions of the wheelchair as shown in Fig. 4.1. Sensors measure the distance from the different obstacles to the chair and then the controller decides the best direction that the wheelchair must follow in order to avoid those obstacles.

The outputs of the neuro-fuzzy controller are the voltages sent to a system that generates a pulse width modulation (PWM) to move the electric motors and the directions in which the wheel will turn. The controller is based on trigonometric neural networks and fuzzy cluster means. It follows a Takagi–Sugeno inference method [2], but instead of using polynomials on the defuzzification process it also uses trigonometric neural networks (T-ANNs). A diagram of the neuro-fuzzy controller is shown in Fig. 4.2.

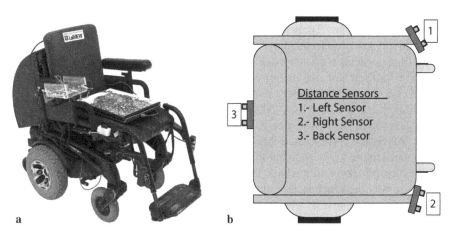

Fig. 4.1 The electric wheelchair with distance sensors

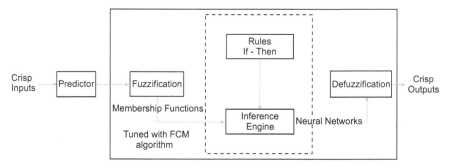

Fig. 4.2 Basic diagram of the neuro-fuzzy controller

4.2.1 Trigonometric Artificial Neural Networks

Consider $f(x)$ to be periodic and integrable in Lebesgue (for continuous and periodic functions (2π) in $[-\pi, \pi]$ or $[0, 2\pi]$; in mathematics, the Lebesgue measure is the standard form to assign a length, area or volume to the subsets of Euclidean space). It must be written as $f \in C * [-\pi, \pi]$ or just $f \in C*$. The Fourier series are associated to f in the point x giving:

$$f(x) \sim \frac{a_0}{2} + \sum_{n=1}^{\infty} (a_n \cos(nx) + b_n \sin(nx)) = \sum_{k=1}^{\infty} A_k(x). \qquad (4.1)$$

The deviation (error) of $f \in C*$ from the Fourier series at the x point or from a trigonometric polynomial of order $\leq n$ is:

$$E_n(f) = \min_{\tau_n} \max_{0 \leq x \leq 2\pi} |f(x) - \tau_n(x)| = \min_{\tau_n} \|f - \tau_n\| . \qquad (4.2)$$

Using Favard sums of f falling in its extreme basic property, give the best approximation for trigonometric polynomials of a class (periodic continuous functions) as follows in (4.3):

$$\|f'\| = \max_{x} |f'(x)| \leq 1. \qquad (4.3)$$

Theorem 4.1. *IF* $f \in C[a,b]$ *and* $\tau_n = P_n$ *is a polynomial of degree* $\delta \leq n$, *THEN* $\lim_{n\to\infty} E_n(f) = 0$.

Using a summation method as in (4.4), where $\mathbf{M}$ is a double matrix of infinite numbers we have:

$$\mathbf{M} = \begin{pmatrix} a_{00} & a_{01} & \cdots & a_{0n} & \cdots \\ a_{10} & a_{11} & & a_{1n} & \cdots \\ \vdots & \vdots & \ddots & \vdots & \\ a_{n0} & a_{n1} & & a_{nn} & \cdots \\ \vdots & \vdots & & \vdots & \end{pmatrix}. \qquad (4.4)$$

For each $\{S_n\}$ sequence the $\{\sigma_n\}$ sequence is associated so that $\sigma_n = \sum_{v=0}^{\infty} a_{nv} S_v$, $n = 0, 1, 2, \ldots$ where the series converge for all n if $\lim_{n\to\infty} \sigma_n = s$. We then say that the sequence $\{S_n\}$ is summable in $\mathbf{M}$ to the limit S. The σ_n are called the linear media of $\{S_n\}$. The equation system $\sigma_n = U \sum a_{nv} S_v$ can be written as $\sigma = T(S)$ and known as a linear transformation . σ_n is also called the transformation of S_n for T. The most important transformations are regulars.

If $y(t)$ is a function in time (a measured signal) and $x(\omega, t)$ is an approximated function (or rebuilt signal) that continuously depends on the vector $\omega \in \Omega$ and of time t, then the problem of decomposition is to find the optimal parameters $\omega* = \left[\omega_1^*, \omega_2^*, \dots, \omega_n^*\right]$ of the approximated function $x(\omega, t) = \sum_{i=1}^{N} \omega_i \Phi_i$, where $\{\Phi_i(t)\}\,(i = 1, 2, \dots, N)$ is a set of basic specific functions. Orthogonal functions are commonly used as basic functions. An important advantage of using orthogonal functions is that when an approximation needs to be improved by increasing the number of basic functions, the ω_i coefficients of the original basic functions remain unchanged. Furthermore, the decomposition of the signal of time in a set of orthogonal functions that are easily generated and defined has many applications in engineering.

Fourier series have been proven to be able to model any periodical signal [3]. For any given signal $f(x)$ it is said to be periodic if $f(x) = f(x + T)$, where T is the fundamental period of the signal. The signal can be modeled using Fourier series :

$$f(x) \sim \frac{a_0}{2} + \sum_{n=1}^{\infty} (a_n \cos(nx) + b_n \sin(nx)) = \sum_{n=1}^{\infty} A_k(x) \tag{4.5}$$

$$a_0 = \frac{1}{T} \int_0^T f(x)\,\mathrm{d}x \tag{4.6}$$

$$a_n = \frac{1}{T} \int_0^T f(x) \cos(n\omega x)\,\mathrm{d}x \tag{4.7}$$

$$b_n = \frac{1}{T} \int_0^T f(x) \sin(n\omega x)\,\mathrm{d}x . \tag{4.8}$$

The trigonometric Fourier series consists of the sum of functions multiplied by a coefficient plus a constant; a neural network can thus be built based on (4.5)–(4.8). Figure 4.3 shows the topology of this network, which is composed of two layers. On the first layer the activation function of the neurons are trigonometric functions. On the second layer the results of the activation functions multiplied by its weights plus a constant are summed. This constant is the mean value of the function; the weights are the coefficients of the Fourier trigonometric series [4].

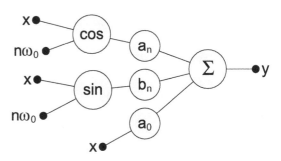

Fig. 4.3 Topology of T-ANNs

The advantages of this topology are that the weights of the network can be computed using analytical methods as a linear equation system. The error of the solution decreases when the number of neurons is augmented, which corresponds to adding more harmonics according to the Fourier series.

To train the network we need to know the available inputs and outputs. The traditional approach to training a network is to assign random values to the weights and then wait for the function to converge using the gradient-descent method. Using this topology the network is trained using the least squares method, fixing a finite number of neurons and arranging the system in a matrix form $Ax = B$. Approximating the function with even functions we use cosines, and if we want to approximate with odd functions we use sines.

Considering the sum of squared differences between the values of the output function, and the ones given by the function $f(x, a_0, \ldots a_n)$ in the corresponding points, we will choose the parameters $a_0, \ldots a_n$ such that the sum will have the minimum value:

$$S(a_0, \ldots a_n) = \sum_{i=1}^{m} [y_i - f(x, a_0, \ldots a_n)]^2 = \min, \qquad (4.9)$$

using cosines

$$S(a_0, \ldots a_n) = \sum_{i=1}^{m} \left[y_i - \left(\frac{1}{2}a_0 + \sum_{k=1}^{\infty} a_k \cos(k\omega_0 x) \right) \right]^2 = \min. \qquad (4.10)$$

This way the problem is reduced to find the parameters $a_0, \ldots, a_n$ for which $S(a_0, \ldots, a_n)$ has a minimum as shown in (4.11) and (4.12).

$$\frac{\partial S}{\partial a_0} - \frac{1}{2} \sum_{i=1}^{m} \left[y_i - \left(\frac{1}{2}a_0 + \sum_{k=1}^{\infty} a_k \cos(k\omega_0 x) \right) \right] = 0 \qquad (4.11)$$

$$\frac{\partial S}{\partial a_p} \sum_{i=1}^{m} \left[y_i - \left(\frac{1}{2}a_0 + \sum_{k=1}^{\infty} a_k \cos(k\omega_0 x) \right) (n\omega_0 x) \right] = 0 \qquad \text{for } p \geq 1. \qquad (4.12)$$

This equation system can be the written in the matrix form $Ax = B$:

$$\begin{bmatrix} \frac{1}{2}m & \cdots & \sum_{i=1}^{m} \cos(p\omega x_i) \\ \frac{1}{2}\sum_{i=1}^{m} \cos(p\omega_0 x_i) & \cdots & \sum_{i=1}^{m} \cos(\omega_0 x_i) \cos(p\omega_0 x_i) \\ \vdots & \ddots & \sum_{i=1}^{m} \cos(\omega_0 x_i) \cos(p\omega x_i) \\ \frac{1}{2}\sum_{i=1}^{m} \cos(p\omega_0 x_i) & \cdots & \sum_{i=1}^{m} \cos^2(p\omega_0 x_i) \end{bmatrix} \begin{bmatrix} a_0 \\ a_1 \\ \vdots \\ a_n \end{bmatrix}$$

$$= \begin{bmatrix} \sum_{i=1}^{m} y_i \\ \sum_{i=1}^{m} y_i \cos(\omega_0 x_i) \\ \vdots \\ \sum_{i=1}^{m} y_i \cos(p\omega_0 x_i) \end{bmatrix}. \qquad (4.13)$$

4.2.1.1 Numerical Example of T-ANNs

The following example shows a numerical approximation made by T-ANNs. Figure 4.4 shows the ANN icon. The front panel and block diagram of the example can also be seen in Fig. 4.5. In the block diagram the code related to training and evaluation of the network is amplified in size.

Four different clusters from representative samples taken by the distance sensors of the wheelchair are included in the program and can be selected by the user. Also, the number of neurons can be varied and the response of the network will change. The code that trains the neural network is shown in Fig. 4.6 and is based on the algorithm previously described. The response of the network trained with different number of neurons is shown in Fig. 4.7.

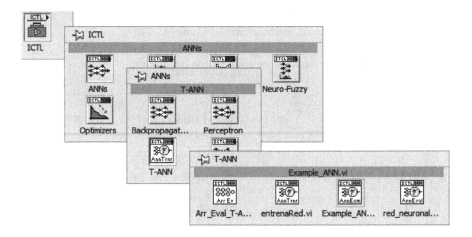

Fig. 4.4 T-ANNs example

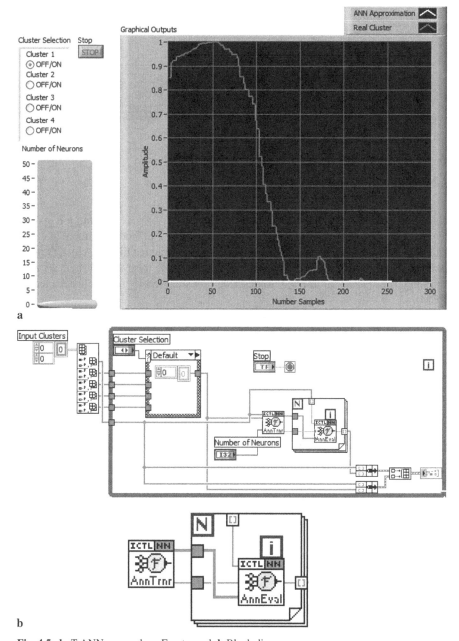

Fig. 4.5a,b T-ANNs example. **a** Front panel. **b** Block diagram

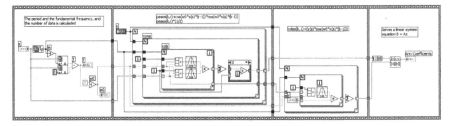

Fig. 4.6 Block diagram of T-ANNs trainer

4.2.2 Fuzzy Cluster Means

Clustering methods split a set of N elements $X = \{x_1, x_2 \ldots, x_n\}$ into a c group denoted $c = \{\mu^1, \mu^2, \ldots \mu^n\}$. Traditional clustering set methods assume that each data vector can belong to one and only one class; in practice though, clusters normally overlap, and some data vectors can belong partially to several clusters. Fuzzy set theory provides a natural way to describe this situation by fuzzy cluster means (FCM).

The fuzzy partition matrices $\mathbf{M} = \{U \in V_{cN} \mid 1, 2, 3\}$, for c classes and N data points were defined by three conditions:

- The first condition: $\forall \; 1 \leq i \leq c \;\; \mu_{ik} \in [0, 1], \;\; 1 \leq k \leq N$.

- The second condition: $\sum_{k=1}^{c} \mu_{ik} = 1 \;\; \forall \; 1 \leq k \leq N$.

- The third condition: $\forall \; 1 \leq i \leq c \;\; 0 < \sum_{k=1}^{c} \mu_{ik} < N$.

The FCM optimum criteria function has the following form $J_m(U, V) = \sum_{i=1}^{c} \sum_{k=1}^{N} \mu_{ik}^{m} d_{ik}^{2}$, where d_{ik} is an inner product norm defined as $d_{ik}^{2} = \|x_k - v_i\|_{\mathbf{A}}^{2}$, $\mathbf{A}$ is a positive definite matrix, and m is the weighting exponent $m \in [1, \infty)$. If m and c parameters are fixed and define sets then (U, V) may be a global minimum for $J_m(U, V)$ only if:

$$\forall \; 1 \leq i \leq c \;\; 1 \leq k \leq N$$

$$u_{ik} = \frac{1}{\sum_{j=1}^{c} \left(\frac{\|x_k - v_i\|}{\|x_k - v_j\|} \right)^{2/(m-1)}} \tag{4.14}$$

$$\forall \; 1 \leq i \leq c$$

$$v_j = \frac{\sum_{k=1}^{N} (u_{ik})^{m} x_k}{\sum_{k=1}^{N} (u_{ik})^{m}} . \tag{4.15}$$

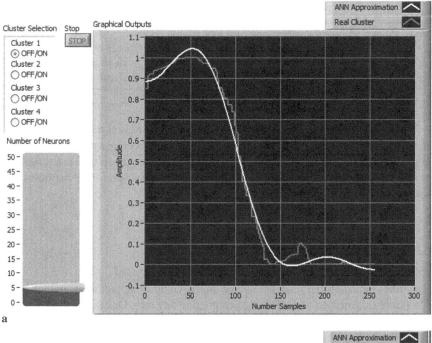

a

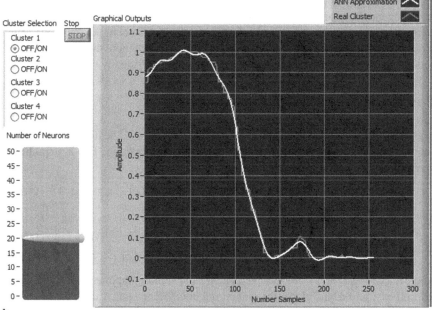

b

Fig. 4.7a,b T-ANN example network. **a** Using 5 neurons. **b** Using 20 neurons

Algorithm 4.1	FCM solution
Step 1	Fix c and m, set $p = 0$ and initialize $U^{(0)}$.
Step 2	Calculate fuzzy centers for each cluster $V^{(p)}$ using (4.15).
Step 3	Update fuzzy partition matrix $\mathbf{U}^{(p)}$ for the pth iteration using (4.14).
Step 4	If $||\mathbf{U}^{(p)} - \mathbf{U}^{(p-1)}|| < \epsilon$ then, $j \leftarrow j + 1$ and return to the *Step 2*.

In this algorithm, the parameter m determines the fuzziness of the clusters; if m is large the cluster is fuzzier. For $m \to 1$ the FCM solution becomes the crisp one, and for $m \to \infty$ the solution is as fuzzy as possible. There is no theoretical reference for the selection of m, and usually $m = 2$ is chosen. After the shapes of the membership functions are fixed, the T-ANNs learn each one of them.

4.2.3 Predictive Method

Sometimes the controller response can be improved by using predictors, which provide future information and allow it to respond in advance. One of the simplest yet most powerful predictors is based on exponential smoothing. A popular approach used is the Holt method.

Exponential smoothing is computationally simple and fast. At the same time, this method can perform well in comparison with other more complex methods. The series used for prediction is considered as a composition of more than one structural component (average and trend) each of which can be individually modeled. We will use series without seasonality in the predictor. Such types of series can be expressed as:

$$y(x) = y_{av}(x) + p y_{tr}(x) + e(x); \quad p = 0, \tag{4.16}$$

where $y(x)$, $y_{av}(x)$, $y_{tr}(x)$, and $e(x)$ are the data, the average, the trend and the error components individually modeled using exponential smoothing. The p-step-ahead prediction [5] is given by:

$$yU * (x + p|k) = y_{av}(x) + p y_{tr}(x). \tag{4.17}$$

The average and the trend components are modeled as:

$$y_{av}(x) = (1 - \alpha) y(x) + \alpha (y_{av}(x - 1) + y_{tr}(k - 1)) \tag{4.18}$$

$$y_{tr}(x) = (1 - \beta) y_{tr}(x - 1) + \beta (y_{av}(x) + y_{av}(x - 1)), \tag{4.19}$$

where $y_{av}(x)$ and $y_{tr}(x)$ are the average and the trend components of the signal, respectively, and α and β are the smoothing coefficients with values in the range $(0, 1)$. The terms y_{av} and y_{tr} can be initialized as:

$$y_{av}(1) = y(1)$$

$$y_{tr}(1) = \frac{(y(1) - y(0)) + (y(2) - y(1))}{2}. \tag{4.20}$$

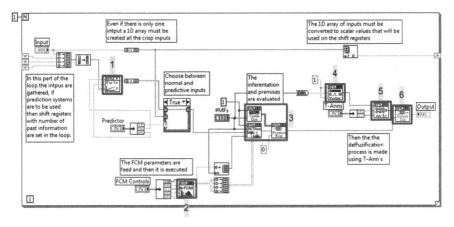

Fig. 4.8 Block diagram of the neuro-fuzzy controller with one input and one output

The execution of the controller (shown in Fig. 4.8) depends on several VIs (more information can be found in [6]), which are explained in the following steps:

1. This is a predictor VI based on exponential smoothing, the coefficients α and β must feed as scalar values. The past and present information must feed in a 1D array with the newest information in the last element of the array.
2. This VI executes the FCM method. The information of the crisp inputs must feed as well as stop conditions for the cycle. The program will return the coefficients of the trigonometric networks, the fundamental frequency and other useful information.
3. These three VIs execute the evaluation of the premises. The first on the top left is generator of the combinations of rules, which depends on the number of inputs and membership functions. The second one on the bottom left evaluates the input membership functions. The last one on the right uses the information on the combinations as well as the evaluated membership functions to obtain the premises of the *IF–THEN* rules.
4. This VI creates a 1D array with the number of rules of the system $\{1, 2, \ldots, n\}$, where n is the number of rules. This array is used in the defuzzification process.
5. This VI evaluates a T-ANN on each of the rules.
6. This VI defuzzifies using the Takagi method with the obtained crisp outputs from the T-ANN.

This version of one input/one output of the controller was modified to have three inputs and four outputs (Fig. 4.9). Each input is fuzzified with four membership functions whose form is defined by the FCM algorithm. The crisp distances gathered by the distance sensors are clustered by FCM and then T-ANNs are trained. As can be seen in Fig. 4.10, the main shapes of the clusters are learned by the neural networks and no main information is lost.

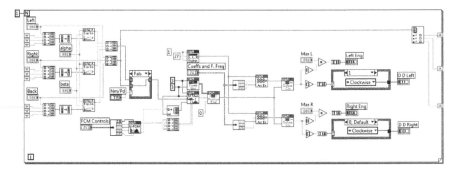

Fig. 4.9 Neuro-fuzzy controller block diagram

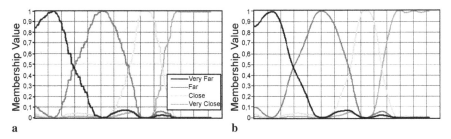

Fig. 4.10a,b Input membership functions. **a** Traditional FCM. **b** Approximated by T-ANNs

With three inputs and four membership functions there are a total of sixty-four rules that can be evaluated. These rules are *IF–THEN* and have the following form: *IF x_1 is μ_{in} AND x_2 is μ_{in} AND x_3 is μ_{in} THEN PWM Left Engine, Direction Left Engine, PWM Right Engine, Direction Right Engine.*

The value of each rule is obtained through the inference method *min* that consists of evaluating the $\mu_{in's}$ and returning the smallest one for each rule. The final system output is obtained by:

$$\text{Output} = \frac{\sum\limits_{i=1}^{r} [\min(\mu_{i1,2,3}) \, NN \, (x_1, x_2, x_3)]}{\sum\limits_{i=1}^{r} \min(\mu_{i1,2,3})}. \tag{4.21}$$

For the direction of the wheel, three states are used: clockwise (1), counterclockwise (−1), and stopped (0). The fuzzy output is rounded to the nearest value and the direction is obtained.

4.2.4 Results Using the Controller

The wheelchair was set on a human-sized chessboard and the pieces where set in a maze as shown in Fig. 4.11, with some of the trajectories described by the chair. The wheelchair always managed to avoid obstacles, but failed to return to the desired

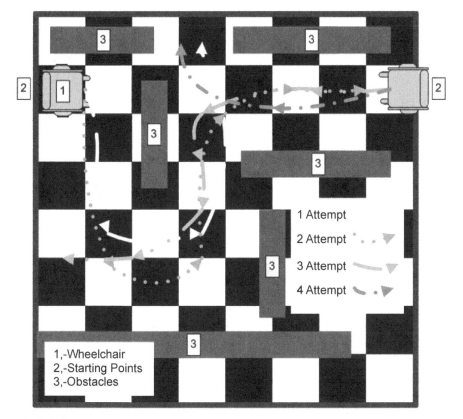

Fig. 4.11 Wheelchair maze and trajectories

direction. It also failed to recognize if the obstacle is a human being or an object and thus, had different behaviors to avoid them.

4.2.5 Controller Enhancements

4.2.5.1 Direction Controller

As seen from the previous results the wheelchair will effectively avoid obstacles but the trajectories that it follows are always different; sometimes it may follow the desired directions but other times it will not. A direction controller can solve this problem. For this we need a sensor to obtain a feedback from the direction of the wheelchair. A compass could be used to sense the direction, either the 1490 (digital) or 1525 (analog) from Images Scientific Instruments [7]. After the electric wheelchair controller avoids an obstacle the compass sensor will give it information to return to the desired direction, as shown in Fig. 4.12.

Fig. 4.12 The wheelchair recovering the direction with the direction controller

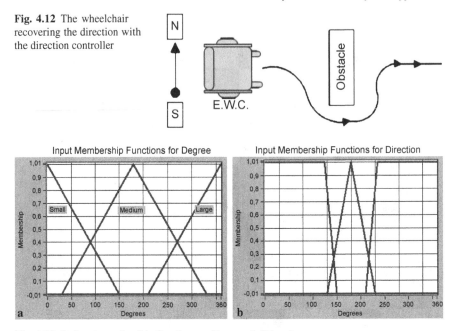

Fig. 4.13a,b Input membership functions. **a** *Degrees*. **b** *Direction*

A fuzzy controller that controls the direction can be used in combination with the obstacle avoidance controller. The directions controller will have as input the difference between the desired and the current direction of the wheelchair. The direction magnitude tells us how many degrees the chair will have to turn, and the sign indicates if it has to be done in one direction or the other. The output is the PWM and the direction that each wheel has to take in order to compensate for that.

Three fuzzifying input membership functions will be used for the degrees and the turning direction, as shown in Fig. 4.13. The range for the degrees is [0, 360], and the turning direction is [−180, 180], also in degrees. The form of the rule is the following: *IF degree is A_{in} AND direction is B_{in} THEN PWM Left Engine, Direction Left Engine, PWM Right Engine, Direction Right Engine.* Table 4.1 shows the rule base with the nine possible combinations of inputs and outputs. The outputs are obtained with the rule consequences using singletons, as illustrated in Fig. 4.14.

The surfaces for the PWM and the direction are shown in Fig. 4.15. For both PWM outputs the surface is the same, while for the direction the surfaces change and completely invert from left to right. This controller will act when the distances recognized by the sensors are *very far*, because the system will have enough space to maneuver and recover the direction that it has to follow, otherwise the obstacle avoidance controller will have control of the wheelchair.

Table 4.1 The *IF–THEN* rules for the direction controller

1. *IF* Degree is Small & Direction is Left *THEN* PWM$_R$ IS Very Few, PWM$_L$ IS Very Few, DIR$_R$ is CCW, DIR$_L$ is CW.
2. *IF* Degree is Small & Direction is Center *THEN* PWM$_R$ IS Very Few, PWM$_L$ IS Very Few, DIR$_R$ is NC, DIR$_L$ is NC.
3. *IF* Degree is Small & Direction is Right *THEN* PWM$_R$ IS Very Few, PWM$_L$ IS Very Few, DIR$_R$ is CW, DIR$_L$ is CCW.
4. *IF* Degree is Medium & Direction is Left *THEN* PWM$_R$ IS Some, PWM$_L$ IS Some, DIR$_R$ is CCW, DIR$_L$ is CW.
5. *IF* Degree is Medium & Direction is Center *THEN* PWM$_R$ IS Some, PWM$_L$ IS Some, DIR$_R$ is NC, DIR$_L$ is NC.
6. *IF* Degree is Medium & Direction is Right *THEN* PWM$_R$ IS Some, PWM$_L$ IS Some, DIR$_R$ is CW, DIR$_L$ is CCW.
7. *IF* Degree is Large & Direction is Left *THEN* PWM$_R$ IS Very Much, PWM$_L$ IS Very Much, DIR$_R$ is CCW, DIR$_L$ is CW.
8. *IF* Degree is Large & Direction is Center *THEN* PWM$_R$ IS Very Much, PWM$_L$ IS Very Much, DIR$_R$ is NC, DIR$_L$ is NC.
9. *IF* Degree is Large & Direction is Right *THEN* PWM$_R$ IS Very Much, PWM$_L$ IS Very Much, DIR$_R$ is CW, DIR$_L$ is CCW.

CCW counterclockwise   *CW* clockwise   *NC* no change

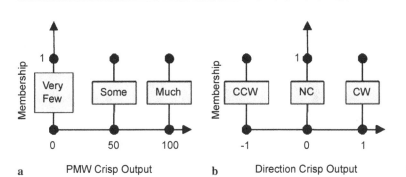

a   PMW Crisp Output   b   Direction Crisp Output

Fig. 4.14 Rule base **a** and output **b** membership functions for the direction controller

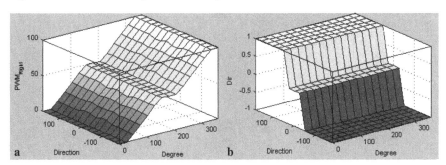

Fig. 4.15a,b Surfaces for outputs. **a** PWM. **b** Direction

4.2.5.2 Obstacle Avoidance Behavior

Cities are not designed with the physically disabled in mind. One of their main concerns is how to go from one point to another. Large cities are becoming more and more crowded so navigating the streets with a wheelchair poses a big challenge.

If temperature and simple shape sensors are installed in the wheelchair (Fig. 4.16) then some kind of behavior can be programmed so that the system can differentiate between a human being and an object. Additionally, the use of a speaker or horn is needed to ask people to move out of the way of the chair.

The proposed behavior is based on a fuzzy controller, which has as input the temperature in degrees of the obstacle and as output the time in seconds the wheelchair will be stopped and a message or horn will be played. It has three triangular fuzzy

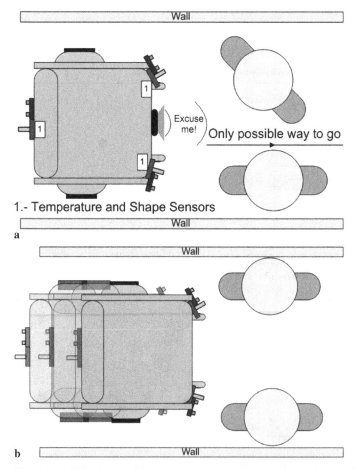

Fig. 4.16a,b Wheelchair with temperature sensors for obstacle avoidance. **a** One possible way to move. **b** Humans detected wheelchair moving forward

Input Membership Functions for Temperature

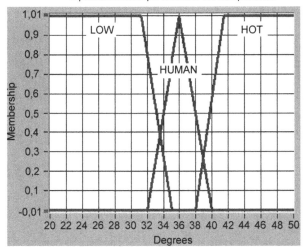

Fig. 4.17 Input membership function for temperature

Table 4.2 The *IF–THEN* rules for the temperature controller

1. *IF* temperature is low *THEN* time is few.
2. *IF* temperature is human *THEN* time is much.
3. *IF* temperature is hot *THEN* time is few.

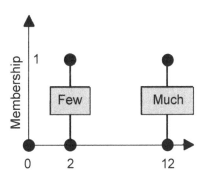

Fig. 4.18 Singleton outputs
for the temperature controller

Time Crisp Output

input membership functions as shown in Fig. 4.17. Table 4.2 shows the *IF–THEN* rules and the output membership functions are two singletons, as seen in Fig. 4.18. The controller response is shown in Fig. 4.19.

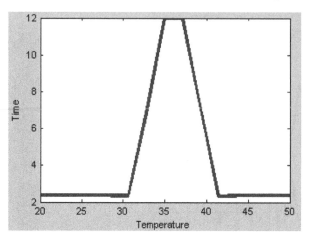

Fig. 4.19 Time controller response

4.3 ANFIS: Adaptive Neuro-fuzzy Inference Systems

Conventional mathematical modeling tools cannot deal with vague or uncertain information. Here is where fuzzy systems using *IF–THEN* rules have the strength and ability to reason as humans, without employing precise and complete information. However, a problem arises as to how to transfer human knowledge to fuzzy systems.

Several proposals have been made, such as the combination of ANNs with fuzzy systems. ANNs have the ability to learn and adapt from experience, thus complementing fuzzy systems. Among the most important techniques is the adaptive neuro-fuzzy inference system (ANFIS) proposed by Jang [8] in 1993, which generates fuzzy *IF–THEN* rule bases and fuzzy membership functions automatically.

ANFIS is based on adaptive networks, which is a super set of feed-forward artificial neural networks with supervised learning capabilities as stated by Jang [8, 9]. It is a topology of nodes directionally connected, where almost all the nodes depend on parameters that are changed according to certain learning rules that will minimize error criteria. The most used learning rule is the gradient-descent method; however, Jang proposed a hybrid learning rule that incorporates least square estimation (LSE).

An adaptive network as shown in Fig. 4.20 is a feed-forward network composed of layers and nodes. Each node performs a function based on incoming signals and parameters associated with the node. No weights are associated with the links, they only indicate flow. Capabilities of the nodes are differentiated by their shape; a square node is adaptive while a round node is fixed.

If we suppose that an adaptive network has L layers and each layer has k nodes, we can denote the node in the ith position of the kth layer by (k, i) and its node function by O_i^k. Thus we can express a node output based on its input signals $\left(O_{\#(k-n)}^{k-n} \right)$ and inherent parameters (a, b, c) as: $O_i^k = O_i^k \left(O_1^{k-1}, \ldots O_{\#(k-1)}^{k-1}, a, b, \ldots \right)$. For

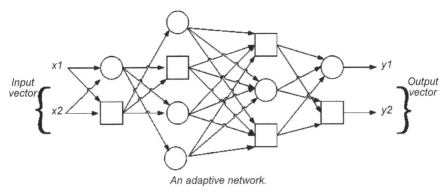

An adaptive network.

Fig. 4.20 Adaptive network

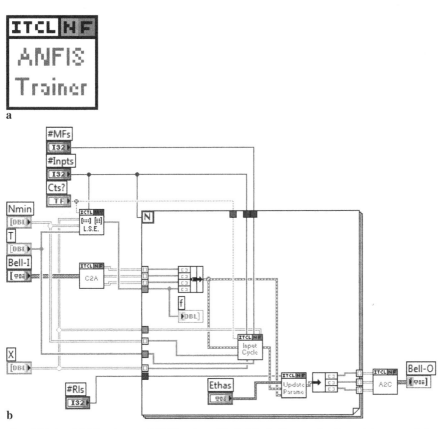

Fig. 4.21a,b ANFIS trainer. **a** ANFIS trainer VIs. **b** Block diagram for the training part of the algorithm

a given training data set P, the error measure (E_p) for the pth $(1 \leq p \leq P)$ sample is defined as the sum of squared error: $E_p = \sum_{m=1}^{L} \left(D_{m,p} - O_{m,p}^{L} \right)^2$, where D is the set of desired output vectors. The gradient-descent rule is based on the error rate. The error rate for the output node is $\frac{\partial E_p}{\partial O_{i,p}^{L}} = -2 \left(D_{i,p} - O_{i,p}^{L} \right)$, and for the internal nodes $\frac{\partial E_p}{\partial O_{i,p}^{k}} = \sum_{m=1}^{k+1} \frac{\partial E_p}{\partial O_{m,p}^{k+1}} \frac{\partial O_{m,p}^{k+1}}{\partial O_{i,p}^{k}}$ for $1 \leq k \leq L - 1$. Finally, the parameter α can be updated with a learning rate of η by $\Delta\alpha = -\eta \frac{\partial E}{\partial \alpha}$.

Jang proposed the hybrid learning rule, where the last layer of the network is trained using LSE. The inputs, parameters and outputs are arranged in matrix form $Ax = B$, and then the unknown parameters x are computed to minimize the squared error given by $\|Ax - B\|$. Because the system is usually overdetermined, there is no exact solution, and thus x is calculated with $x = \left(A^T A \right)^{-1} A^T B$. This step is usually computed once at the beginning of the iteration process so the computational burden is lowered, but it can also be computed at each iteration.

The VIs **anfis_trainer-bell.vi** and **anfis_trainer-triangular.vi** execute the training algorithm for bell and triangular input membership functions. The programs first adjust the parameters of the last layer using LSE then perform the gradient-descent method on the parameters of the inner layers. Figure 4.21 shows the icon and the block diagram for the bell functions case.

4.3.1 ANFIS Topology

Assuming a fuzzy inference system with two inputs x, y and one output z. A first-order Sugeno system is shown in Fig. 4.22a and the corresponding ANFIS topology is illustrated in Fig. 4.22b. A sample from one rule is the following: *Rule n: IF x is A_n and y is B_n THEN $f_n = p_n x + q_n y + r_n$.*

Layer 1

Every node in this layer is adaptive with a function $O_{1,i} = \mu_{A_i}(x)$, where x (or y) is the input to the ith node and A_i (or B_i) is the linguistic label with the node. $O_{1,i}$ is the membership function of A_i or B_i, usually a bell-shaped function defined by the function:

$$\mu_{A_i}(x) = \frac{1}{1 + \left[\left(\frac{x-c_i}{a_i} \right)^2 \right]^{b_i}}, \tag{4.22}$$

where a_i, b_i, c_i are the parameters that define the form of the bell. Parameters in this layer are referred to as premise parameters. Figure 4.23 shows the VI for the evaluation of multiple inputs with multiple number of fuzzy Jang bell functions.

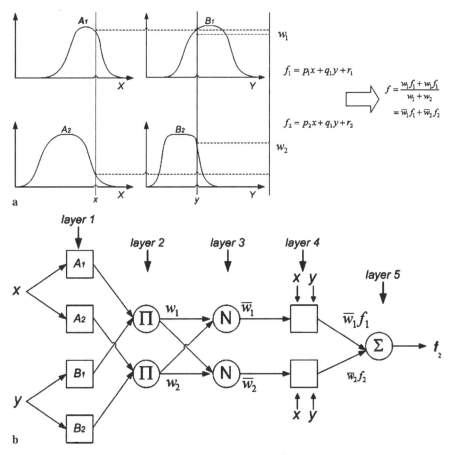

Fig. 4.22a,b Fuzzy and ANFIS type 3. **a** Type-3 fuzzy reasoning, corresponding to a Sugeno system. **b** Equivalent ANFIS type-3

Fig. 4.23 MIMO VI for fuzzy
Jang bell functions

Layer 2

Nodes in this layer are fixed, whose output is the product of all incoming signals, representing the firing strength of the rule. Any T-norm operator performing a fuzzy *AND* can be used as the node function $O_{2,i} = \omega_i = \mu_{A_i}(x)\mu_{B_i}(x)$.

a

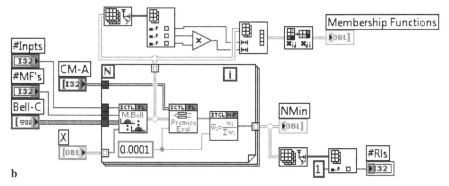

b

Fig. 4.24a,b ANFIS execution up to layer 3. **a** VI ANFIS layer 3. **b** Block diagram of the VI

Layer 3

This node normalizes the firing strength (N); they are of fixed form and calculate the ratio of the ith rule firing strength to the sum of firing strength of all rules by:

$$O_{3,i} = \overline{\omega_i} = \frac{\omega_i}{\sum_r \omega_r}. \tag{4.23}$$

There is a VI that executes the algorithm to train an ANFIS up to layer 3, as shown in Fig. 4.24. The program receives the number of inputs of the system, the number of membership functions (all inputs have the same number of membership functions) of the inputs, and the parameters of the functions. It is then evaluated up to the premise part.

Layer 4

These nodes are called the consequence parameters; they are of the adaptive class. The node function is $O_{4,i} = \overline{\omega}_i \, (px_i + q_i y + r_i)$, where (p_i, q_i, r_i) are the parameters of the node.

Layer 5

This is a fixed node (Σ), which computes the overall output of the system, as the summation of all signals as:

$$\text{Output} = O_{5,1} = \sum_i \bar{\omega}_i f_i = \frac{\sum_i \omega_i f_i}{\sum_i \omega_i}. \tag{4.24}$$

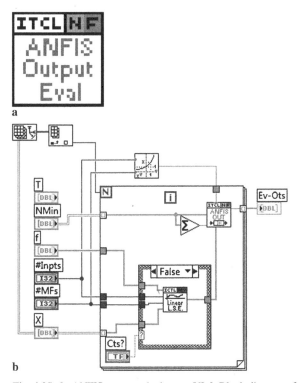

Fig. 4.25a,b ANFIS output calculator. **a** VI. **b** Block diagram of the VI

$$O_i^1 = \mu A_i(x)$$

$$\mu A_i(x) = \frac{1}{1 + \left\{ \left(\frac{x - c_i}{a_i} \right)^2 \right\} bi}$$

$$\mu A_i(x) = \exp \left\{ + \left(\frac{x - c_i}{a_i} \right)^2 \right\}$$

$$w_i = \mu A_i(x) {}^* \mu B_i(y), \quad i = 1, 2.$$

$$\overline{w}_i = \frac{w_i}{w_1 + w_2}, \quad i = 1, 2$$

$$O_i^4 = \overline{w}_i f_i = \overline{w}_i (p_i x + q_i y + r_i)$$

$$O_i^5 = \text{overall output} = \sum_i \overline{w}_i f_i = \frac{\sum_i \overline{w}_i f_i}{\sum_i w_i}$$

Fig. 4.26 ANFIS architecture

The output of the system (Fig. 4.25) can be calculated using the **anfis_evaluator.vi**, which performs the execution of layers 4 and 5. The ANFIS architecture equations for this system are shown in Fig. 4.26. This network has the functionality of a Sugeno model. The conversion from Sugeno ANFIS (type-3) to Tsukamoto (type-1) is straightforward, as illustrated in Fig. 4.27.

For the Mamdani reasoning system (type-2) using max–min composition , a corresponding ANFIS can be constructed if discrete approximations are used to replace the integrals in the centroid defuzzification scheme. However the resulting ANFIS

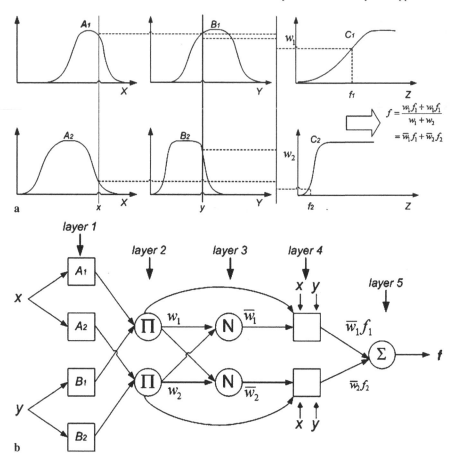

Fig. 4.27a,b Fuzzy and ANFIS type-1. **a** Type-1 fuzzy reasoning. **b** Equivalent ANFIS type-1

is much more complex than either the Sugeno or Tsukamoto type and does not necessarily imply better learning capabilities.

A slightly more complex Sugeno ANFIS is shown in Fig. 4.28. Three membership functions are associated with each input, so the input space is partitioned into nine regions. Figure 4.28b illustrates that each of these regions is governed by a fuzzy *IF–THEN* rule. The premise part of the rules defines a fuzzy region, while the consequent part specifies the output within the region.

Example 4.1. On a control field, an identification system is one of its targets. Suppose that we have a machine, which tends to a sinc function (sampling function). In order to understand this example, let us suppose that the behavior of that machine is related to the sinc function in the interval [0, 200]. Using a hybrid learning method, train an ANFIS to approximate this neuro-fuzzy system into the machinery behav-

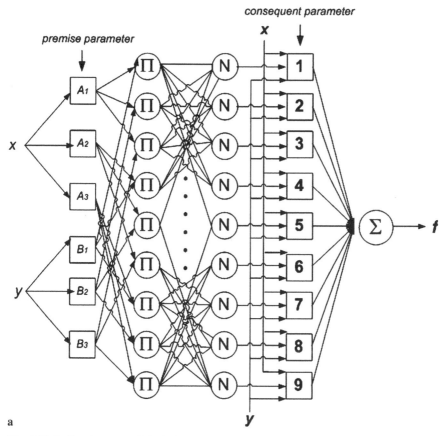

Fig. 4.28a,b Two-input type-3 ANFIS with 9 rules. **a** ANFIS topology. **b** Corresponding to fuzzy subspaces

ior. Engineers require five bell membership functions (Gaussian functions) to solve this problem and constant functions at layer 4. Use a 0.05 learning rate value for any parameter in bell functions.

Solution. The Intelligent Control Toolkit implements this example following the path ICTL ≫ Neuro-Fuzzy ≫ ANFIS ≫ **Example_ANFIS-Bell.vi**. We will first describe the front panel (Fig. 4.29). At the top-left of the window is the *Cycle Parameters* option. In this section, there are the *MFs* values that control the number of membership functions that ANFIS will have in layer 1. Also, the *Stop Conditions* are *Min E*, referring to the minimum error accepted, and *MaxI*, referring to the maximum number of iterations performed by the training procedure. The *Training function* options are used to select the function that we want to approximate (sinc function in this example). If *Train?* is selected then the ANFIS initializes its learning procedure; if *Cts?* is selected then layer 4 are constant functions, or else functions

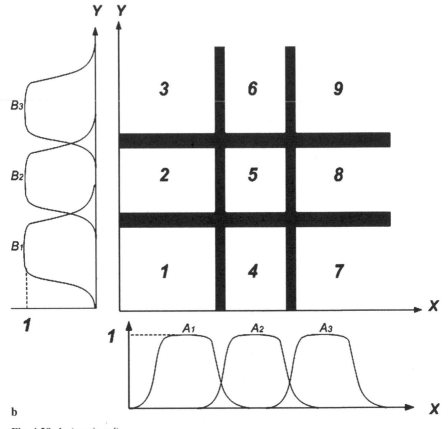

Fig. 4.28a,b (continued)

at this layer are adapted and trained by least squares error . *Ethas* is a cluster with the learning rate values for each of the parameters that the ANFIS has in its adaptive nodes. In the last case, membership functions are Gaussian or bell functions that have three parameters a, b and c; this is the reason why the *Ethas* cluster has *ctea, cteb* and *ctec* representing the three learning rates.

Below the *Cycle Parameters* are the *Input/Output* options. The 2D array *Inputs* refers to the input values of the ANFIS, and the *Trainer* array contains the elements that ANFIS will try to approximate (sinc function). *#Inps* returns the number of inputs and *Iterations* shows the actual iteration. Finally, there is the *Best Parameters Found So Far* section. In this block of options are the *BestB* that contains the best bell function representation by its parameters, *BErr* returns the value of the minimum error found so far, and *Best f* refers to the constants that represent the linear function that produces *BErr.*

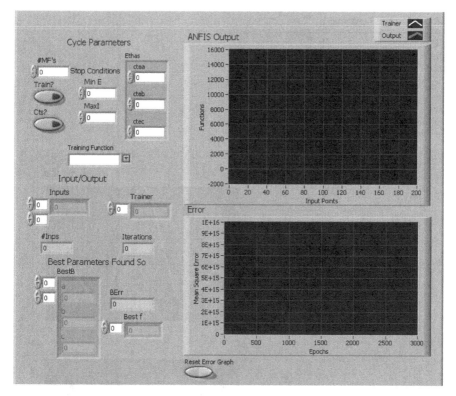

Fig. 4.29 Front panel of the ANFIS example using bell membership functions

On the right side of Fig. 4.29 is the *ANFIS Output* graph that represents the *Trainer* function (sinc function) and the actual *Output* of the trained ANFIS. The *Error* graph represents the error values at each epoch.

As we know, the bell function in the range [0, 1] is represented mathematically as:

$$f(x) = \frac{1}{\left(\frac{x-c}{a}\right)^{2b} + 1} . \tag{4.25}$$

Then, this VI will adapt the parameters a, b and c from the above equation. The minimum error and maximum iterations are proposed in Table 4.3. In this example, the VI delimits the sinc function in the interval [0, 200]. Running the program, we can look at the behavior of the training procedure as in Figs. 4.30–4.32. Remember to switch on the *Train?* and *Cts?* buttons. ☐

Example 4.2. We want to control a 12V DC motor in a fan determined by some ambient conditions. If the temperature is less than 25 °C, then the fan is switched off. If the temperature is greater than 35 °C, then the fan has to run as fast as possible.

Table 4.3 ANFIS example 1

MFs	5
Min E	1E–5
MaxI	10 000
Ctea	0.05
Cteb	0.05
Ctec	0.05
Training function	Sinc

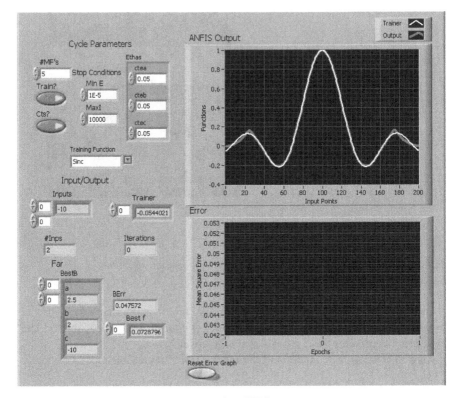

Fig. 4.30 Initial step in the training procedure for ANFIS

If the temperature is between 25 °C and 35 °C, then the fan has to follow a logistic function description. In this way, we know that the velocity of rotation in a DC motor is proportional to the voltage supplied. Then, the logistic function is an approximation of the voltage that we want to have depending on the degrees of the environment. The function is described by (4.26).

$$f(x) = \frac{1}{e^{-ax} + 1}, \quad \forall a \in \Re. \tag{4.26}$$

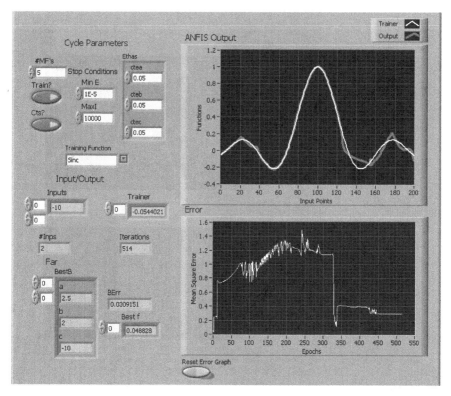

Fig. 4.31 Training procedure for ANFIS at 514 epochs

A simple analysis offers that the range of the logistic function is $[0, 12]$, and for limiting the domain of the function, suppose an interval $[0, 60]$. Select $a = 2.5$. Using the hybrid learning method, train an ANFIS system selecting four triangular membership functions with learning rates for all parameters equal to 0.01. Determine if this number of membership functions is optimal; otherwise propose the optimal number.

Solution. Following the path ICTL $\gg$ Neuro-Fuzzy $\gg$ ANFIS $\gg$ **Example_ANFIS-Triangular.vi**. As in Example 4.1, this VI is very similar except that the adaptive parameters come from triangular membership functions. Remember that a triangular membership function is defined by three parameters: a means the initial position of the function, b is the value at which the function takes the value 1, and c is the parameter in which the function finishes.

We need to modify the block diagram. First, add a *Case Before* in the *Case Structure* as shown in Fig. 4.33. Label this new case as "Logistic." Then, access ICTL $\gg$ ANNs $\gg$ Perceptron $\gg$ Transfer F. $\gg$ **logistic.vi**. This function needs *Input* values coming from the vector node (see Figs. 4.33 and 4.34) and a 2.5 constant is placed in the *alpha* connector.

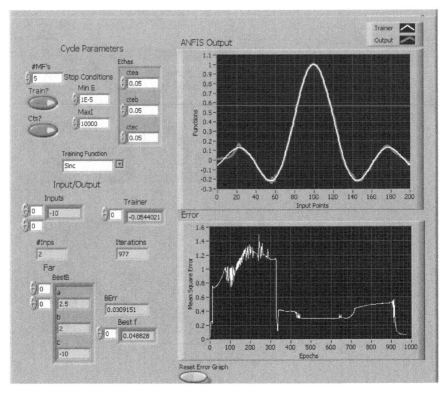

Fig. 4.32 Training procedure for ANFIS at 977 epochs

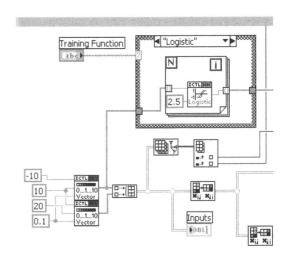

Fig. 4.33 Case structure for
the logistic function

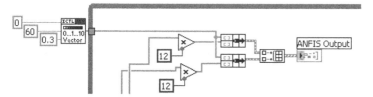

Fig. 4.34 Block diagram showing the corrections in the ANFIS graph

Table 4.4 ANFIS example 2

MFs	4
Min E	1E–5
MaxI	10 000
Ctea	0.01
Cteb	0.01
Ctec	0.01
Training function	Logistic

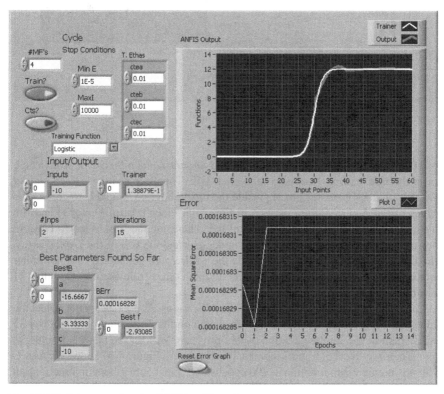

Fig. 4.35 Training procedure for ANFIS at 15 epochs

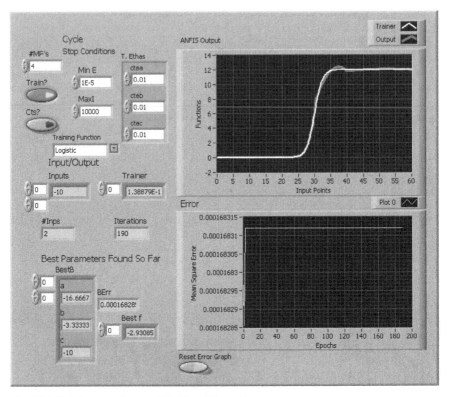

Fig. 4.36 Training procedure for ANFIS at 190 epochs

After that, add a new item in the *Training Function Combo Box* at the front panel. Label the new item as "Logistic" and click OK. Then, replace the *ANFIS Output Graph* with a *XY Graph*. Looking inside the block diagram, we have to correct the values of the *ANFIS Output* as seen in Fig. 4.35. Place a **rampVector.vi** and run this node from 0 to 60 with a stepsize of 0.3. These numbers are selected because they are the domain of the temperature in degrees and the size of the *Trainer* array. The first orange line (the top one inside the *while-loop*) connected to a multiplier comes from the *Trainer* line and the second line comes from the *Ev-Ots* output of the **anfis_evaluator.vi**.

Then, the VI is available for use with the indications. At the front panel, select the values shown in Table 4.4. Remember to switch on the *Train?* button. Figures 4.35 and 4.36 show the implementation of that program. We can see that the training is poor. Then, we select 5, 6 and 7 membership functions. Figure 4.37 shows the results with this membership function at 200 epochs. We see at 5 membership functions an error of 4.9E–4, at 6 an error of 1.67E–5, and at 7 an error of 1.6E–4. We determine that the optimal number of membership functions is 6.   □

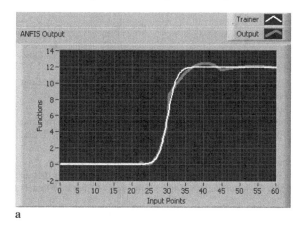

a

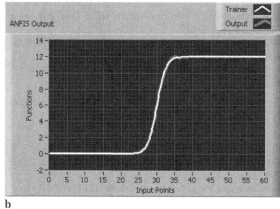

b

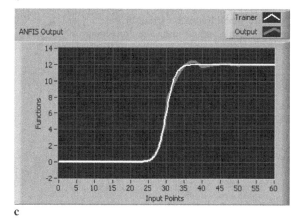

c

Fig. 4.37a–c ANFIS responses at 200 epochs. **a** With 5 membership functions. **b** With 6 membership functions. **c** With 7 membership functions

References

1. Ponce P, et al. (2007) Neuro-fuzzy controller using LabVIEW. Proceedings of 10th ISC Conference, IASTED, Cambridge, MA, 19–21 Nov 2007
2. Takagi T, Sugeno M (1998) Fuzzy identification of systems and its application to modeling and control. IEEE Trans Syst Man Cyber 15:116–132
3. Fourier J (2003) The analytical theory of heat. Dover, Mineola, NY
4. Ponce P, et al. (2006) A novel neuro-fuzzy controller based on both trigonometric series and fuzzy clusters. Proceedings of IEEE International Conference on Industrial Technology, India, 15–17 Dec 2006
5. Kanjilal PP (1995) Adaptive prediction and predictive control. Short Run, Exeter, UK
6. Ramirez-Figueroa FD, Mendez-Cisneros D (2007) Neuro-fuzzy navigation system for mobile robots. Dissertation, Electronics and Communications Engineering, Tecnológico de Monterrey, México, May 22 2007
7. Images Scientific Instrumentation (2009) http://www.imagesco.com. Accessed on 22 Feb 2009
8. Jang J-SR (1993) ANFIS: adaptive network-based inference system. IEEE Trans Syst Man Cyber 23(3): 665–685
9. Jang J-SR, Sun C-T, Mizutani E (1997) Neuro-fuzzy and soft computing: a computational approach to learning and machine intelligence. Prentice Hall, New York
10. ITESM-CCM Team (2009) Electric wheelchair presented in NIWEEK 2009 Austin Texas

Futher Reading

Jang JSR (1992) Self-Learning Fuzzy Controller Based on Temporal Back-Propagation. IEEE Trans. on Neural Networks, 3:714–723

Jang JSR (1993) ANFIS: Adaptive-Network-based Fuzzy Inference Systems. IEEE Trans. on Systems, Man, and Cybernetics, 23:665–685

Jang JSR, Sun CT (1993) Functional Equivalence Between Radial Basis Function Networks and Fuzzy Inference Systems. IEEE Trans. on Neural Networks, 4:156–159

Jang JSR, Sun CT (1995) Neuro-Fuzzy Modeling and Control. The Proceedings of the IEEE, 83:378–406a

Chapter 5
Genetic Algorithms and Genetic Programming

5.1 Introduction

In this chapter we introduce powerful optimization techniques based on evolutionary computation. The techniques mimic natural selection and the way genetics works. Genetic algorithms were first proposed by J. Holland in the 1960s. Today, they are mainly used as a search technique to find approximate solutions to different kinds of problems. In intelligent control (IC) they are mostly used as an optimization technique to find minimums or maximums of complex equations, or quasi-optimal solutions in short periods of time.

N. Cramer later proposed genetic programming in 1985, which is another kind of evolutionary computation algorithm with string bases in genetic algorithms (GA). The difference basically is that in GA strings of bits representing chromosomes are evolved, whereas in genetic programming the whole structure of a computer program is evolved by the algorithm. Due to this structure, genetic programming can manage problems that are harder to manipulate by GAs. Genetic programming has being used in IC optimize the sets of rules on fuzzy and neuro-fuzzy controllers.

5.1.1 Evolutionary Computation

Evolutionary computation represents a powerful search and optimization paradigm. The metaphor underlying evolutionary computation is a biological one, that of natural selection and genetics. A large variety of evolutionary computational models have been proposed and studied. These models are usually referred to as evolutionary algorithms. Their main characteristic is the intensive use of randomness and genetic-inspired operations to evolve a set of solutions.

Evolutionary algorithms involve selection, recombination, random variation and competition of the individuals in a population of adequately represented potential solutions. These candidate solutions to a certain problem are referred to as chromosomes or individuals. Several kinds of representations exist such as bit string,

P. Ponce-Cruz, F. D. Ramirez-Figueroa, *Intelligent Control Systems with LabVIEW™*
© Springer 2010

real-component vectors, pairs of real-component vectors, matrices, trees, tree-like hierarchies, parse trees, general graphs, and permutations.

In the 1950s and 1960s several computer scientists started to study evolutionary systems with the idea that evolution could be applied to solve engineering problems. The idea in all the systems was to evolve a population of candidates to solve problems, using operators inspired by natural genetic variations and natural selection.

In the 1960s, I. Rechenberg introduced evolution strategies that he used to optimize real-valued parameters for several devices. This idea was further developed by H.P. Schwefel in the 1970s. L. Fogel, A. Owens and M. Walsh in 1966 developed *evolutionary programming*, a technique in which the functions to be optimized are represented as a finite-state machine, which are evolved by randomly mutating their state-transition diagrams and selecting the fittest. Evolutionary programming, evolution strategies and GAs form the backbone of the field of evolutionary computation.

GAs were invented by J. Holland in the 1960s at the University of Michigan. His original intention was to understand the principles of adaptive systems. The goal was not to design algorithms to solve specific problems, but rather to formally study the phenomenon of adaptation as it occurs in nature and to develop ways in which the mechanisms of natural adaptation might be ported to computer systems. In 1975 he presented GAs as an abstraction of biological evolution in the book *Adaptation in Natural and Artificial Systems*.

Simple biological models based on the notion of survival of the best or fittest were considered to design robust adaptive systems. Holland's method evolves a population of candidate solutions. The chromosomes are binary strings and the search operations are typically crossover, mutation, and (very seldom) inversion. Chromosomes are evaluated by using a fitness function.

In recent years there has been an increase in interaction among researchers studying different methods and the boundaries between them have broken down to some extent. Today the term GA may be very far from Holland's original concept.

5.2 Industrial Applications

GAs have been used to optimize several industrial processes and applications. F. Wang and others designed and optimized the power stage of an industrial motor drive using GAs at the Virginia Polytechnic Institute and State University at Virginia in 2006 [1]. They analyzed the major blocks of the power electronics that drive an industrial motor and created an optimization program that uses a GA engine. This can be used as verification and practicing tools for engineers.

D.-H. Cho presented a paper in 1999 [2] that used a niching GA to design an induction motor for electric vehicles. Sometimes a motor created to be of the highest efficiency will perform at a lower level because there are several factors that were not considered when it was designed, like ease of manufacture, maintenance, reliability, among others. Cho managed to find an alternative method to optimize the design of induction motors.

GAs have also been used to create schedules in semiconductor manufacturing systems. S. Cavalieri and others [3] proposed a method to increase the efficiency of dispatching, which is incredibly complex. This technique was applied to a semiconductor manufacture plant. The algorithm guarantees that the solution is obtained in a time that is compatible with on-line scheduling. They claim to have increased the efficiency by 70%.

More recently V. Colla and his team presented a paper [4] where they compare traditional approaches, and GAs are used to optimize the parameters of the models. These models are often designed from theoretical consideration and later adapted to fit experimental data collected from the real application. From the results presented, the GA clearly outperforms the other optimization methods and fits better with the complexity of the model. Moreover, it provides more flexibility, as it does not require the computation of many quantities of the model.

5.3 Biological Terminology

All living organisms consist of cells that contain the same set of one or more *chromosomes* serving as a blueprint. Chromosomes can be divided into *genes*, which are functional blocks of DNA. The different options for genes are *alleles*. Each gene is located at a particular *locus* (position) on the chromosome. Multiple chromosomes and or the complete collection of genetic material are called the organism's *genome*. A *genotype* refers to the particular set of genes contained in a genome.

In GAs a *chromosome* refers to an individual in the population, which is often encoded as a string or an array of bits. Most applications of GAs employ haploid individuals, which are single-chromosome individuals.

5.3.1 Search Spaces and Fitness

The term "search space" refers to some collection of candidates to a problem and some notion of "distance" between candidate solutions. GAs assume that the best candidates from different regions of the search space can be combined via crossover, to produce high-quality offspring of the parents. "Fitness landscape" is another important concept; evolution causes populations to move along landscapes in particular ways and adaptation can be seen as the movement toward local peaks.

5.3.2 Encoding and Decoding

In a typical application of GAs the genetic characteristics are encoded into bits of strings. The encoding is done to keep those characteristics in the environment. If we want to optimize the function $f(x) = x^2$ with $0 \leq x < 32$, the parameter of the search space is x and is called the phenotype in an evolutionary algorithm. In

Table 5.1 Chromosome encoded information

Decimal number	Binary encoded
5	00101
20	10100
7	01011

GAs the phenotypes are usually converted to genotypes with a coding procedure. By knowing the range of x we can represent it with a suitable binary string. The chromosome should contain information about the solution, also known as encoding (Table 5.1).

Although each bit in the chromosome can represent a characteristic in the solution here we are only representing the numbers in a binary way. There are several types of encoding, which depend heavily on the problem, for example, permutation encoding can be used in ordering problems, whereas floating-point encoding is very useful for numeric optimization.

5.4 Genetic Algorithm Stages

There are different forms of GAs, however it can be said that most methods labeled as GAs have at least the following common elements: population of chromosomes, selection, crossover and mutation (Fig. 5.1). Another element rarely used called inversion is only vaguely used in newer methods. A common application of a GA is the optimization of functions, where the goal is to find the global maximum or minimum.

A GA [5] can be divided into four main stages:

- *Initialization*. The initialization of the necessary elements to start the algorithm.
- *Selection*. This operation selects chromosomes in the population for reproduction by means of evaluating them in the fitness function. The fitter the chromosome, the more times it will be selected.

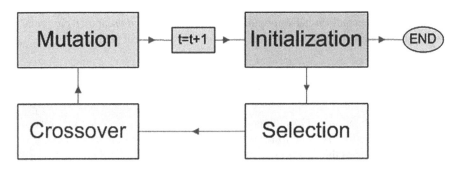

Fig. 5.1 GA main stages

- *Crossover.* Two individuals are selected and then a random point is selected and the parents are cut, then their tails are crossed. Take as an example 100110 and 111001: the 3 position from left to right is selected, they are crossed, and the offspring is 100001, 111110.
- *Mutation.* A gene, usually represented by a bit is randomly complemented in a chromosome, the possibility of this happening is very low because the population can fall into chaotic disorder.

These stages will be explained in more detail in the following sections.

5.4.1 Initialization

In this stage (shown in Fig. 5.2) the initial individuals are generated, and the constants and functions are also initiated, as shown in Table 5.2.

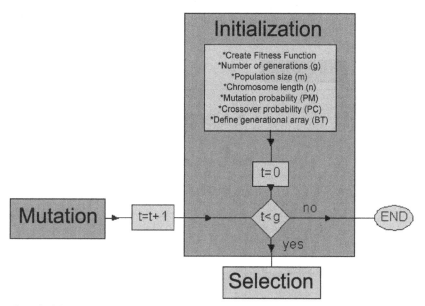

Fig. 5.2 GA initialization stage

Table 5.2 GA initialization parameters

Parameter	Description
g	The number of generations of the GA.
m	Size of the population.
n	The length of the string that represents each individual: $s = \{0, 1\}^n$. The strings are binary and have a constant length.
PC	The probability of crossing of 2 individuals.
PM	The probability of mutation of every gen.

5.4.2 Selection

A careful selection of the individuals must be performed because the domination of a single high-fit individual may sometimes mislead the search process. There are several methods that will help avoid this problem, where individual effectiveness plays a very negligible role in selection. There are several selection methods like scaling transformation and rank-based, tournaments, and probabilistic procedures.

By scaling we mean that we can modify the fitness function values as required to avoid the problems connected with proportional selection. It may be static or dynamic; in the latter, the scaling mechanism is reconsidered for each generation.

Rank-based selection mechanisms are focused on the rank ordering of the fitness of the individuals. The individuals in a population of n size are ranked according to their suitability for search purposes. Ranking selection is natural for those situations in which it is easier to assign subjective scores to the problem solutions rather than to specify an exact objective function.

Tournament selection was proposed by Goldberg in 1991, and is a very popular ranking method of performing selection. It implies that two or more individuals

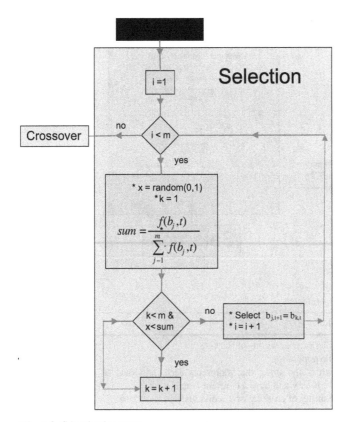

Fig. 5.3 GA selection stage

compete for selection. The tournament may be done with or without reinsertion of competing individuals into the original population.

Finally the roulette-wheel selection process is another popular method used as selection stage, where the fittest individuals have a higher probability to be selected. In this method individuals are assigned a probability to be selected, then a random number is calculated and the probability of individuals is accumulated. Once that value of the random number is reached, the individual presently used is the one that is selected.

However, in order to perform selection it is necessary to introduce a measure of the performance of individuals. By selection we aim to maximize the performance of individuals. Figure 5.3 shows the diagram of this stage.

Fitness Function

As we already mentioned selection methods need a tool to measure the performance of individuals. The search must concentrate on regions of the search space where the best individuals are located. This concentration accomplishes the exploitation of the best solutions already found, which is exactly the purpose of selection. For selection purposes a performance value is associated with each individual in the current population, and represents the *fitness* of the function.

A *fitness function* is usually used to measure explicitly the performance of chromosomes, although in some cases the fitness can be measured only in an implicit way, using information about the performance of systems. Chromosomes in a GA take the form of bit strings; they can be seen as points in the search space. This population is processed and updated by the GA, which is mainly driven by a fitness function, a mathematical function, a problem or in general, a certain task where the population has to be evaluated.

5.4.3 Crossover

In order to increase population diversity, other operators are used such as the crossover operation. By perturbing and recombining the existent individuals, the search operators allow the search process to explore the neighboring regions or to reach further promising regions.

Crossover operations achieve the recombination of the selected individuals by combining segments belonging to chromosomes corresponding to parents. Figure 5.4 shows a diagram of the crossover stage, which creates an information exchange between the parent chromosomes. Thus the descendent obtained will possess features from both parents.

The role of recombination is to act as an impetus to the search progress and to ensure the exploration of the search space. Various crossover operations have been proposed, and here we will explain the most employed variant used in binary encoded frameworks: the one-point crossover. The crossover probability (CP) is

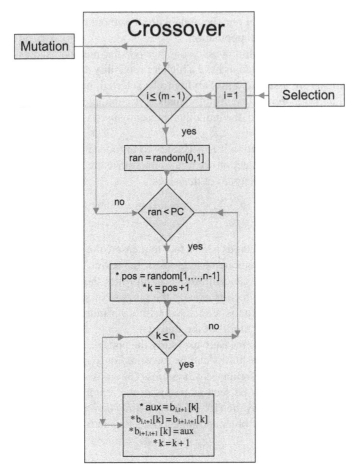

Fig. 5.4 GA crossover stage

compared with a random number between $(0,1)$ and with this, it is determined if is going to be crossover or not. When a crossover is made, the positions in which the parents are going to be cut in a random position are then interchanged.

5.4.4 Mutation

In classical genetics, mutation is identified by an altered phenotype, and in molecular genetics mutation refers to any alternation of a segment of DNA. Spontaneous mutagenesis is normally not adaptive, and mutations normally do not provide a selective advantage. Changes may destroy the genome structure, where other changes tend to create and integrate new functions.

Changes representing a selective disadvantage occur considerably more often and can affect life processes in various degrees. The extreme situation leads to lethality. Often the alteration of the chromosome remains without immediate consequences on life processes. This kind of mutations is called *natural* or *silent*. Neutral mutations however may play an evolutionary role.

Within the framework of binary encoding mutation is considered the second most important genetic operator. The effect of this operator is to change a single position (gene) within a chromosome. If it were not for mutation, other individuals could not

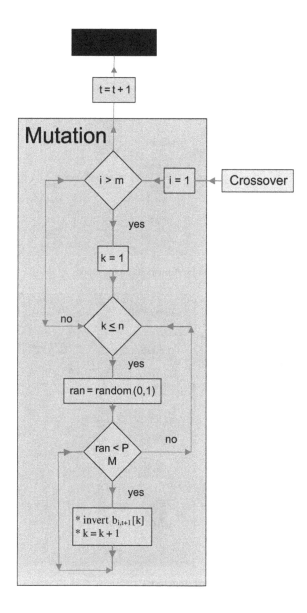

Fig. 5.5 Mutation stage

be generated through other mechanisms, which are then introduced to the population. The mutation operator assures that a full range of individuals is available to the search.

One of the simplest executions of mutation is when the mutation probability (MP) is compared with a random number between (0,1). If it is going to be a mutation, a randomly chosen bit of the string is inverted. A diagram showing this stage is in Fig. 5.5.

Example 5.1. This is an example of a GA using the Intelligent Control Toolkit for LabVIEW (ICTL). A base algorithm created for searching the maximum and minimum in the $f(x) = x^2$ function will be explained. This program is included as

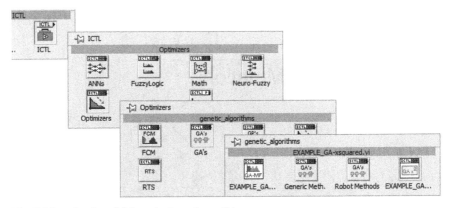

Fig. 5.6 Localization of GA methods on the toolkit

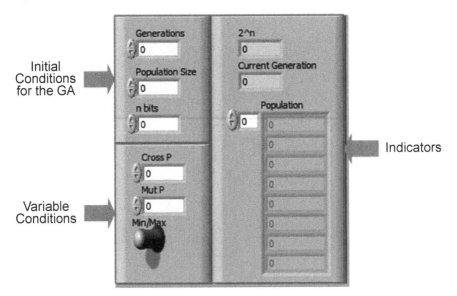

Fig. 5.7 GA-X squared front panel

Table 5.3 Initial conditions and variables for the x^2 example

Variable	Description
Generations	The number of repetitions of the algorithm.
Population Size	The number of individuals per generation, for example 6.
n bits	In this case it is how many bits will be in the binary string that represents the individuals, for example, if the individuals will be numbers from 0 to 31, then $n = 5$, because $25 - 1 = 31$.
Cross P	The crossover probability if $CP = 1$ every individual selected will get combined with other selected individuals.
Mut P	Is the mutation probability and should be small; 0.001 is a good value.
Min/Max	Allows us to select if we want to minimize or maximize the function.

a toolkit example but we will explain the development in detail. Figure 5.6 shows where the GA methods can be found in the ICTL. First we build the front panel where we will have the controls and indicators of all the variables.

Figure 5.7 shows an image of how to build the front panel. The initial conditions and variables that we will include are shown in Table 5.3. Now we need to start building the code of our program. First, we need to generate an initial population before we start the algorithm, therefore we create a series of random numbers, depending on the initial conditions. The code is shown in Fig. 5.8.

Basically, a series of random numbers are created, where the top is given by the number of bits used, and later they are transformed into chains of bits. We also need a decoding function and a fitness function. The decoding will convert bits to numbers again and the fitness function will raise those numbers to the second power, as shown in Fig. 5.9.

Now we need to perform selection, crossover, and mutation, the basic operations of a GA. As shown in Fig. 5.10, we find the GA methods in the path Optimizers ≫ GAs Palette ≫ Generic Methods. Our decoded and fitness-evaluated individuals will be fed to the selection method, later we will start a loop where the selected

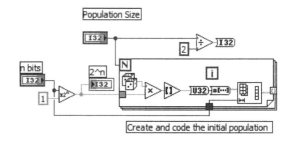

Fig. 5.8 Creation and coding of initial population

Fig. 5.9 Decoding and fitness function

Fig. 5.10 GA methods included in the ICTL

ga_crossover.vi ga_mutation.vi ga_selection-... ga_selection.vi

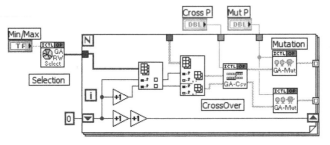

Fig. 5.11 Selection, crossover, and mutation stages code

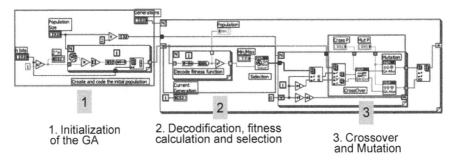

1. Initialization of the GA

2. Decodification, fitness calculation and selection

3. Crossover and Mutation

Fig. 5.12 Complete block diagram of the GA-X squared example

individuals will be crossed over and mutated (Fig. 5.11). From this we also see that we can perform operations in parallel, e. g., the mutation, thus allowing us to increase the operation time of the program.

The complete code is found in Fig. 5.12. Now we can set the initial conditions and run our program and see that after 100 generations the maximum of the function $f(x) = x^2$ for $0 \leq x \leq 31$ has been found; Fig. 5.13 shows some results. With this simple but powerful example we can apply GAs to other applications; the key will be in the coding and fitness function.   □

5.5 Genetic Algorithms and Traditional Search Methods

There are at least three meanings of *search* in which we should be interested:

1. *Search stored data.* The problem to be solved here is to efficiently retrieve stored information; an example can be search in computer memory. This can be applied to enormous databases, which nowadays can be found in many forms on the

Fig. 5.13 GA-X squared
showing some results

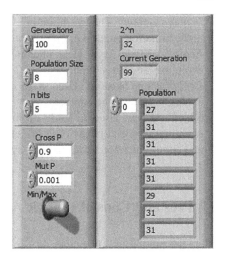

Internet. For example, what is the best way to search for hotels in a particular city? *Binary search* is one method for efficiently finding a desired record in a database.

2. *Search paths and goals.* This search form can be seen as the movements to go from a desired initial state to a final state, like the shortest path in a maze.

3. *Search for solutions.* This is a more general search form of the search for paths and goals. This happens because a path through a search tree can be encoded as a candidate solution. The idea is to efficiently find a solution to a problem in a large space of candidate solutions. These are the most common problems to which GAs are applied.

5.6 Applications of Genetic Algorithms

Even though the foundations of GAs are very simple, certain variations have been used in a large number of scientific and engineering applications. Some of them are given below:

• *Optimization.* The optimization of mathematical functions, combinatorial and numerical problems such as circuit layout and job-shop schedule.

• *Automatic programming.* Used to evolve computer programs, and to design computational structures.

• *Machine learning.* Classification and prediction tasks, such as weather or protein structure. GAs have been used to evolve aspects of particular machine learning systems such as the weights in neural networks, rules for learning classifier systems, sensors and robots.

• *Economics.* Development of bidding strategies, model processes of innovation, strategies, emergence of markets.

- *Immune systems.* The evolution of evolutionary time in multi-gene families, somatic mutation, natural immune systems.
- *Ecology.* Model ecological phenomena such as symbiosis, host-parasite co-evolution, and resource flow.
- *Evolution and learning.* Used in the study of how individual learning and species evolutions affect one another.
- *Social systems.* Used to study evolutionary aspects of social systems, like the evolution of social behavior in insect colonies, and the evolution of cooperation and communication in multi-agent systems.

5.7 Pros and Cons of Genetic Algorithms

There are numerous advantages to using a GA, such as not depending on analytical knowledge, robustness, and intuitive operation. All of these characteristics have made GAs strong candidates in search and optimization problems. However, there are also several disadvantages to using GAs that have made researchers turn to other search techniques, such as:

- Probabilistic.
- Expensive in computational resources.
- Prone to premature convergence.
- Difficult to encode a problem in the form of a chromosome.

There are several alternatives that have been found, especially due to the difficulty of encoding the problems. Messy GAs and genetic programming are two techniques that are based on the framework of GAs.

5.8 Selecting Genetic Algorithm Methods

The representation, recombination, mutation, and selection are complex balances between exploitation and exploration. It is a matter of precision to maintain this balance, thus there are several key factors that can help us correctly choose the encoding technique.

Encoding. This is a key issue with most evolutionary algorithms, whether to choose a suitable encoding scheme, which could be binary, floating-point, or grammatical. On the one hand, Holland supports the idea of a genome with a smaller number of alleles with long strings, rather than a numeric scheme with a larger number of alleles but short (floating-point) strings. On the other hand, M. Mitchell points out that for real-world applications it is more natural to use decimals or symbolic representation [6]. The conclusion from Z. Michalewicz about this is that a floating-point scheme is faster, more consistent between runs, and can provide a higher precision for large-domain applications.

Operator choice. There are a few general guidelines to follow when choosing an operator. Many advantages can be obtained from a two-point crossover operation, by reducing the disruption of schemas with a long length. The choice of the mutation depends heavily on the application, a practical alternative is an adaptive mutation parameterized within the genome.

Elitism. Another common technique in many GAs is to retain the best candidate in each generation and pass it to the next; this can give a significant boost to the performance of the GA, although with the risk of reducing diversity.

5.9 Messy Genetic Algorithm

There are several approaches with regard to the modification of certain aspects of GAs, with the aim of improving their performance. Messy GAs were proposed by D. Goldberg and co-workers in 1989, where they used variable-length binary encodings of chromosomes.

Each gene of the chromosome is represented by a pair (position and value), ensuring the adaptation of the algorithm to a larger variety of situations. Moreover this representation prevents the problems generated by recombination. A messy GA adapts its representation to the problem being solved.

The operators used in this algorithm are generalizations of standard genetic operators that use binary encoding. The main disadvantage against fixed-length representations is that they lack dynamic variability; thus, by limiting the string length the search space is limited. To overcome this, variable-length representations allow us to deal with partial information or to use contradictory information.

Messy encoding. Chromosome length is variable and genes may be arranged in any order (messy), where the last characteristic is the one that gives the algorithm its name. Each gene is represented by a pair of numbers. The first component is a natural number that encodes the gene location; the second number represents the gene value, which usually is either 0 or 1.

Example 5.2. Considering the binary encoded chromosome $x = (01101)$, which can be transformed into the following sequence: $x' = ((1, 0), (2, 1), (3, 1), (4, 0), (5, 1))$. The meaning of this chromosome does not change if the pairs are arranged in a different order, for instance the following chromosome: $x'' = ((2, 1), (3, 1), (1, 0), (4, 0), (5, 1))$. □

Incompleteness and ambiguity. As chromosomes have a flexible structure, we may consider missing one or more genes, which is called an *underspecified* string. This allows us to encode and deal with incomplete information. The opposite situation is overspecification, which occurs when a string contains multiple pairs for the same gene creating redundant or even contradictory genes.

To deal with overspecification, certain rules can be applied such as the tie-breaking mechanism that essentially says "first-come, first served," so that only the first of the repeated genes is taken into consideration. To deal with underspecified strings several possibilities exist, like looking for the complete chromosome that is

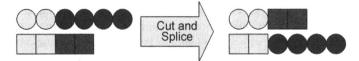

Fig. 5.14 Cut and splice operation

closest to the underspecified string. Another way is to try to approximate the absent value or to identify the probability p that the missing gene has the value of 1; if the value is 0 then the probability will be $(1 - p)$. Another way to do it is by using competitive templates, considered as locally optimal strings.

Crossover. The classical n-point crossover is replaced by the *cut-and-splice* opera-
tor, which acts very similarly to a one-point crossover. Two parents are cut in two and the resulting substrings are recombined. The position for the crossover is chosen with a probability that is uniform to the string length. The difference is that the crossover points are independent from the two parents.

The *splice* operation concatenates the substrings obtained through cutting, where Fig. 5.14 shows this operation. There is no restriction regarding the way in which substrings are combined. The tie-breaking rule along with competitive templates is used to handle overspecified and underspecified strings.

Messy GAs have provided results for difficult problems. In the case of deceptive functions, messy GAs perform better than simple GAs, usually finding the best solu-
tion. An important computational problem within messy algorithms is the dimension of the search space, i. e., since large chromosomes may appear, the dimension could be very high. The search space size is a polynomial. In parallel implementations, the search time is reduced and it is logarithmic with respect to the number of variables of the search space.

5.10 Optimization of Fuzzy Systems Using Genetic Algorithms

A brief explanation on how GAs can be applied to optimize the performance of a fuzzy controller is given in this section.

5.10.1 Coding Whole Fuzzy Partitions

There is always knowledge of the desired configuration, for example, the number of clusters and the labels for each one, where a natural order of the fuzzy sets can be established. By including the proper constraints, the initial conditions can be preserved while reducing the number of degrees of freedom in order to maintain the interpretability of a fuzzy system. Thus, we can encode a whole fuzzy partition as shown in (5.1), where σ is the upper boundary for the size of the offset for example $\sigma = (b - a)/2$. Figure 5.15 shows the coded triangular membership function.

$$C_{n,[0,\partial]}(x_1) \rightarrow C_{n,[0,\partial]}(x_2 - x_1) \cdots C_{n,[0,\partial]}(x_{2N-2} - x_{2N-3}) . \qquad (5.1)$$

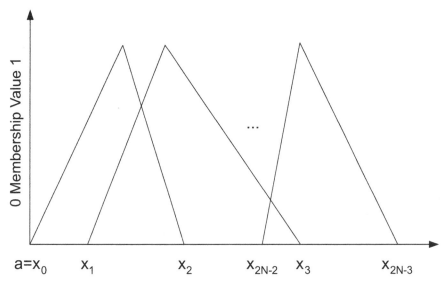

Fig. 5.15 Coded triangular membership functions

5.10.2 Standard Fitness Functions

We define the fitness function as a function that will minimize the distance between a set of representative inputs and outputs, and the values computed by the next function, where the sum of quadratic errors is calculated using (5.2):

$$f(v) = \sum_{i=1}^{k} (F(v, x_i) - y_i)^2. \tag{5.2}$$

Here, $F(v, x_i)$ is the function that computes the output with respect to the parameter vector, (x_i, y_i) is the sample data given as a list of couples, $1 \le i \le k$, and k is the number of samples.

5.10.3 Coding Rule Bases

So far we have explained in which way membership functions can be encoded to be optimized with GAs, but if we find a proper method to encode rule bases into a string of fixed lengths we can apply the previously explained GAs to optimize them without modification.

We must assume that the number of linguistic values of all linguistic variables are finite. A rule base is represented as a list for one input variable, as a matrix for two variables, and as a tensor in the case of more than two variables. This rule can be represented by a matrix as shown in Fig. 5.16. Consider the rule base of the form

Fig. 5.16 Example decision
table

	$B_1 \ldots B_{N2}$
A_1	$C_{1,1} \ldots C_{i,N2}$
.	. . .
.	. . .
.	. . .
A_{N1}	$C_{N1,1} \ldots C_{N1,N2}$

in (5.3):

$$\text{IF } x_1 \text{ is } A_i \text{ AND } x_2 \text{ is } B_j \text{ THEN } C_{i,j} . \tag{5.3}$$

We can now assign indices to the linguistic values associated with elements of the set $\{C_{i,j}\}$. We can later write the decision table as an integer string, and convert those numbers to bits, where the previously mentioned GAs are perfectly suitable to optimize the rule base.

5.11 An Application of the ICTL for the Optimization of a Navigation System for Mobile Robots

A navigation system based on Bluetooth technology was designed for controlling a quadruped robot in unknown environments which has ultrasonic sensor as inputs for avoiding static and dynamic obstacles. The robot Zil I is controled by a fuzzy logic controller Sugeno Type, which is shown in Fig. 2.24 and Zil I is shown in Fig. 5.17, the form of the membership functions for the inputs are triangular and the

Fig. 5.17 Robot Zil I was controlled by a Fuzzy Logic Controller adjusted using Genetic Algo-rithmsC

Fig. 5.18 Triangular membership functions

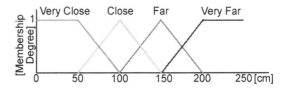

initial membership function's domain and shape are shown in Fig. 5.18. The block diagram of the fuzzy controller is shown in Sect. 5.11 ICTL for the Optimization of a Navigation System for Mobile Robots.

The navigation system is based on a Takagi–Sugeno controller, which is shown in Fig. 5.17. The form of the membership function is triangular, and the initial limits are shown in Fig. 5.18. The block diagram of the fuzzy controller is shown in Fig. 5.19. Based on the scheme of optimization of fuzzy systems using GAs, the fuzzy controller was optimized. Some initial individuals where created using expert knowledge and others were randomly created. In Fig. 5.20 we see the block diagram of the GA.

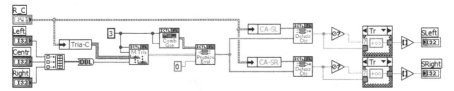

Fig. 5.19 Block diagram of the Takagi–Sugeno controller

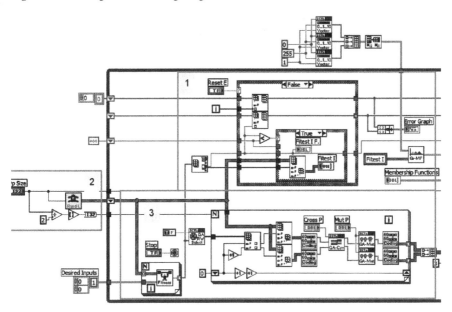

Fig. 5.20 Block diagram of the GA

Inspecting the block diagram we find that the GA created previously, used for the optimization of the $f(x) = x^2$ function, remains the same. The things that change here are the coding and decoding functions, as well as the fitness function. There is also some code used to store the best individuals. After running the program for a while, the form of the membership functions will vary from our initial guess, as shown in Fig. 5.21, and it will find an optimized solution that will fit the constrains set by the human expert knowledge and the requirements for the application. The solutions are shown in Fig. 5.22.

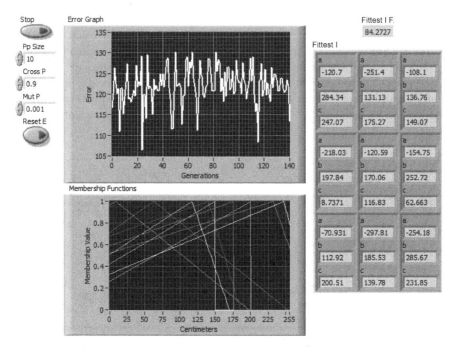

Fig. 5.21 Results shown by the GA after some generations

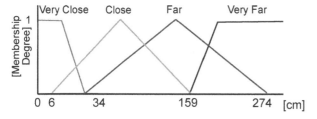

Fig. 5.22 Optimized membership functions

5.12 Genetic Programming Background

Evolution is mostly determined by natural selection, which can be described as individuals competing for all kinds of resources in the environment. The better the individuals, the more likely they will propagate their genetic material. Asexual reproduction creates individuals identical to their parents; this is done by the encoding of genetic information. Sexual reproduction produces offspring that contain a combination of information from each parent, and is achieved by combining and reordering the chromosomes of both parents.

Evolutionary algorithms have been applied to many problems such as optimization, machine learning, operation research, bioinformatics and social systems, among many others. Most of the time the mathematical function that describes the system is not known and the parameters that are known are found through simulation.

Genetic programming, evolutionary programming, evolution strategies and GAs are usually grouped under the term evolutionary computation, because they all share the same base of simulating the evolution of individual structures. This process depends on the way that performance is perceived by the individual structures as defined by the problem.

Genetic programming deals with the problem of automatic programming; the structures that are being evolved are computers programs. The process of problem solving is regarded as a search in the space of computer programs, where genetic programming provides a method for searching the fittest program with respect to a problem. Genetic programming may be considered a form of *program discovery.*

5.12.1 Genetic Programming Definition

Genetic programming is a technique to automatically create a working computer program from a high-level statement of the problem. This is achieved by genetically breeding a population of computer programs using the principles of Darwinian natural selection and biologically inspired operators. It is the extension of evolutionary learning into the space of computer programs.

The individual population members are not fixed-length character strings that encode possible solutions of the problem, they are programs that when executed are the candidate solutions to the problem. These programs are represented as trees. There are other important components of the algorithm called terminal and function sets. The terminal set consists of variables and constants. The function sets are the connectors and operators that relate the constants and variables.

Individuals evolved from genetic programming are program structures of variable sizes. A user-defined language with appropriate operators, variables, and constants may be defined for the particular problem to be solved. This way programs will be generated with an appropriate syntax and the program search space limited to feasible solutions.

5.12.2 Historical Background

A.M. Turing in 1950, considered the fact that genetic or evolutionary searches could automatically develop intelligent computer programs, like chess player programs and other general purpose intelligent machines. Later in 1980, Smith proposed a classifier system that could find good poker playing strategies using variable-sized strings that could represent the strategies. In 1985, Cramer considered a tree structure as a program representation in a genotype. The method uses tree structures and subtree crossover in the evolutionary process.

Genetic programming was first proposed by Cramer in 1985 [7], and further developed by Koza [8], as an alternative to fixed-length evolutionary algorithms by introducing trees of different shapes and sizes. The symbols used to create these structures are more varied than zeros and ones used in GAs. The individuals are represented by genotype/phenotype forms, which make them non-linear. They are more like protein molecules in their complex and unique hierarchical representation. Although parse trees are capable of exhibiting a great variety of functionalities, they are highly constrained due to the form of tree, the branches are the ones that are modified.

5.13 Industrial Applications

Some interesting applications of genetic programming in the industry are mentioned here. In 2006 J.U. Dolinsky and others [9] presented a paper with an application of genetic programming to the calibration of industrial robots. They state that most of the proposed methods address the calibration problem by establishing models followed by indirect and often ill-conditioned numeric parameter identification. They proposed an inverse static kinematic calibration technique based on genetic programming, used to establish and identify model parameters.

Another application is the use of genetic programming for drug discovery in the pharmaceutical industry [10]. W.B. Langdon and S.K. Barrett employed genetic programming while working in conjunction with GlaxoSmithKline (GSK). They were invited to predict biochemical activity using their favorite machine learning technique. Their genetic programming was the best of 12 tested, which marginally improved the existing system of GSK.

5.14 Advantages of Evolutionary Algorithms

Probably the greatest advantage of evolutionary algorithms is their ability to address problems for which there are no human experts. Although human expertise is to be used when available, it has proven less than adequate for automating problem-solving routines.

A primary advantage of this kind of algorithm is that they are simple to represent. They can be modeled as a difference equation $x[t+1] = s(r(x[t]))$, which can be understood as: $x[t]$ is the population at time t under the representation x, is the random variation operator and s is the selection operator.

The representation does not affect the performance of the algorithm, in contrast with other numerical techniques, which are biased on continuous values or constrained sets. They offer a framework to easily incorporate known knowledge of the problem, which could yield in a more efficient exploration and response of the search space.

Evolutionary algorithms can be combined with simple or complex traditional optimization techniques. Most of the time the solution can be evaluated in parallel, and only the selection must be processed serially. This is an advantage over other optimization techniques like tabu search and simulated annealing. Evolutionary algorithms can be used to adapt solutions to changing circumstances, because traditional methods are not robust to dynamic changes and often require a restart to provide the solution.

5.15 Genetic Programming Algorithm

In 1992, J.R. Koza developed a variation of GAs that is able to automate the generation of computer programs [8]. Evolutionary algorithms, also known as evolutionary computing, are the general principles of natural evolution that can be applied to completely artificial environments. GAs and genetic programming are types of evolutionary computing.

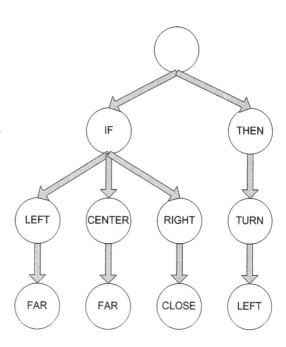

Fig. 5.23 Tree representation of a rule

Genetic programming is a computing method, which provides a system with the possibility of generating optimized programs or computer codes. In genetic programming *IF-THEN* rules are coded into individuals, which often are represented as trees. For example, a rule for a wheeled robot may be *IF left is far AND center if far AND right is close THEN turn left*. This rule is represented as a tree in Fig. 5.23.

According to W. Banzhaf "genetic programming, shall include systems that constitute or contain explicit references to programs (executable code) or to programming language expressions."

5.15.1 Length

In GAs the length of the chromosome is fixed, which can restrict the algorithm to a non-optimal region of the problem in search space. Because of the tree representation, genetic programming can create chromosomes of almost any length.

5.16 Genetic Programming Stages

Genetic programming uses four steps to solve problems:

1. Generate an initial population of random compositions of functions and terminals of the problem (computer programs).
2. Execute each program in the population and assign it a fitness value according to how well it solves the problem.
3. Create a new population of computer programs:

   a. Copy the best existing programs.
   b. Create new programs by mutation.
   c. Create new computer programs by crossover.

4. The best computer program that appeared in any generation, the best-so-far solution, is designated the result of genetic programming [8].

Just like in GAs, in genetic programming the stages are initialization, selection, crossover, and mutation.

5.16.1 Initialization

There are two methods for creating the initial population in a genetic programming system:

1. *Full* selects nodes from only the function set until a node is at a specified maximum depth.
2. *Grow* randomly selects nodes from the function and terminal set, which are added to a new individual.

5.16.2 Fitness

It could be the case that a function to be optimized is available, and we will just need to program it. But for many problems it is not easy to define an objective function. In such a case we may use a set of training examples and define the fitness as an error-based function. These training examples should describe the behavior of the system as a set of input/output relations.

Considering a training set of k examples we may have (x_i, y_i), $i = 1, \ldots, k$, where x_i is the input of the ith training sample and y_i is the corresponding output. The set should be sufficiently large to provide a basis for evaluating programs over a number of different significant situations.

The fitness function may also be defined as the total sum of squared errors; it has the property of decreasing the importance of small deviations from the target outputs. If we define the error as $e_i = (y_i - o_i)^2$ where y_i is the desired output and o_i the actual output, then the fitness will be defined as $\sum_{i=1}^{k} e_k = \sum_{i=1}^{k} (y_i - o_i)^2$. The fitness function may also be scaled, thus allowing amplification of certain differences.

5.16.3 Selection

Selection operators within genetic programming are not specific; the problem under consideration imposes a particular choice. The choice of the most appropriate selection operator is one of the most difficult problems, because generally this choice is problem-dependent. However, the most-used method for selecting individuals in genetic programming is tournament selection, because it does not require a centralized fitness comparison between all individuals. The best individuals of the generation are selected.

5.16.4 Crossover

The form of the recombination operators depends on the representation of individuals, but we will restrict ourselves to tree-structured representations. An elegant and rather straightforward recombination operator acting on two parents swaps a subtree of one parent with a subtree of the other parent.

There is a method proposed by H. Iba and H. Garis to detect regularities in the tree program structure and to use them as guidance for the crossover operator. The method assigns a performance value to a subtree, which is used to select the crossover points. Thus, the crossover operator learns to choose good sites for crossover.

Simple crossover operation. In a random position two trees interchange their branches, but it should be in a way such that syntactic correctness is maintained. Each offspring individual will pass to the selection process of the next generation. In Fig. 5.24 a representation of a crossover is shown.

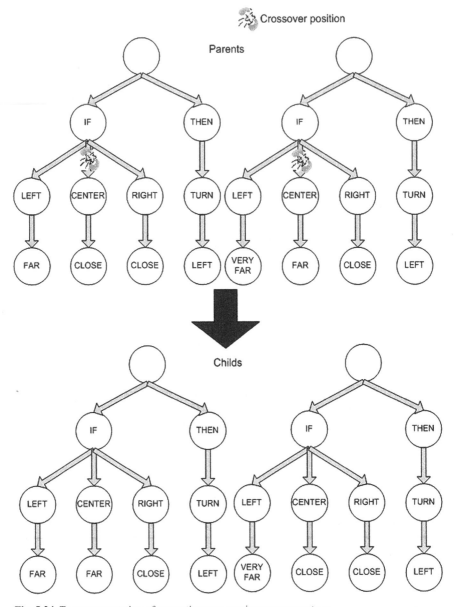

Fig. 5.24 Tree representation of a genetic programming crossover stage

5.16.5 Mutation

There are several mutation techniques proposed for genetic programming. An example is the mutation of tree-structured programs; here the mutation is applied to a single program tree to generate an offspring. If our program is linearly repre-

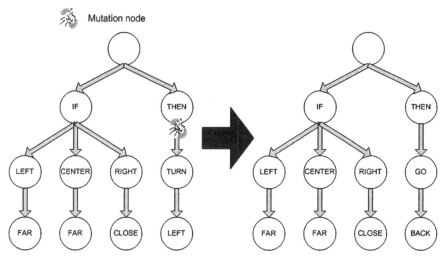

Fig. 5.25 Tree representation of a genetic programming mutation stage

sented, then the mutation operator selects an instruction from the individual chosen for mutation. Then, this selected instruction is randomly perturbed, or is changed to another instruction randomly chosen from a pool of instructions.

The usual strategy is to complete the offspring population with a crossover operation. On this kind of population the mutation is applied with a specific mutation probability. A different strategy considers a separate application of crossover and mutation. In this case it seems to be emphasized with respect to the previous, standard technology.

In genetic programming, the generated individuals are selected with a very low probability of being mutated. When an individual is mutated, one of its nodes is selected randomly and then the current subtree at that point is replaced with a new randomly generated subtree. It is important to state that just as in biological mutation, in genetic programming mutation the genotype may not change but the resulting genotype could be completely different (Fig. 5.25).

5.17 Variations of Genetic Programming

Several variations of genetic programming can be found in the literature. Some of them are linear genetic programming, a variant that acts on linear genomes rather than trees; gene expression programming, where the genotype (a linear chromosome) and the phenotype (expression trees) are different entities that form an indivisible whole; multi-expression programming encodes several solutions into a chromosome; Cartesian genetic programming uses a network of nodes (indexed graph) to achieve an input-to-output mapping; and traceless genetic programming, which does not store explicitly the evolved computer programs, and is useful when the relation between the input and output is not important.

5.18 Genetic Programming in Data Modeling

The main purpose of evolutionary algorithms is to imitate natural selection and evolution, allowing the most efficient individuals to reproduce more often. Genetic programming is very similar to GAs; the main difference is that genetic programming uses different coding of potential solutions. By using knowledge from great amounts of data collected from different places, we can discover patterns and represent them in a way that humans can understand them.

By mathematical modeling we understand that certain equations fit some numerical data. It is used in a variety of scientific problems, where the theoretical foundations are not enough to give answers to experiments. Sometimes using traditional methods is not enough because these methods assume a specific form of model. Genetic programming allows the search for a good model in a different and more "intelligent" way, and can be used to solve highly complex, non-linear, chaotic problems.

5.19 Genetic Programming Using the ICTL

Here we will continue with the optimization example of fuzzy systems. As we have mentioned previously, a Takagi–Sugeno is the core controller of a navigation system that maneuvers the movements of a quadruped robot in order to avoid obstacles. We have previously optimized the form of the membership functions using GAs and now we will evolve the form of the rules and modify their operators. The rules for the controllers used in the quadruped robot have the form of (5.4):

$$IF \text{ Left is } ALi \text{ } Conn \text{ Central is } ACi \text{ } Conn \text{ Right is } ARi \text{ } THEN \text{ } SLeft, SRight,$$
(5.4)

where $ALi = ACi = ARi$ are the number of fuzzifying membership functions for the inputs, in this case they are the same, and $Conn$ is the operation to be performed. There are four options: (1) min, (2) max, (3) $product$, and (4) sum. $SLeft, SRight$ are the speeds used to control the movements of the robot.

Genetic programming will be applied using the following convention to code-decode the individuals:

- 3 bits that will help us determine if the set is complemented or not. A bi-dimensional array must be generated with the same form of the CM-A of the **input-combinator-generator.vi** that is used to evaluate the different sets of the possible rule combinations. The first dimension contains the number of rules, the second the number of inputs to the system.
- 2 bits are used for the premise evaluation. The premises of the next connection operation are used: (0) min, (1) max, (2) $product$, and (3) sum.
- 10 bits are used to obtain the outputs of the rule, 5 for each output to obtain a constant between 0 and 31.

Table 5.4 Individual rule coding for genetic programming ICTL example

Three bits IS, IS NOT			Two bits Conn		10 bits for rule output									
0	1	2	3	4	5	6	7	8	9	10	11	12	13	14

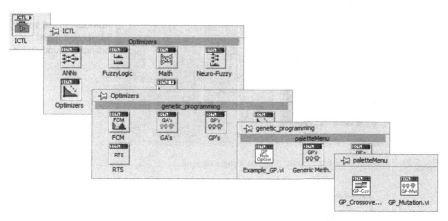

Fig. 5.26 Localization of genetic programming methods on the ICTL

The rule in bits is shown in Table 5.4. As shown in Fig. 5.26, the methods for genetic programming are found at Optimizers ≫ GPs ≫ Generic Methods. An individual contains 27 of these rules; the **initial_population.vi** initializes a population with random individuals. The code is shown in Fig. 5.27. Fixed individuals may be added based on human expert knowledge.

Fig. 5.27 Block diagram for the initialization of random individuals

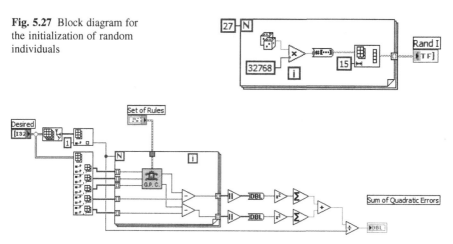

Fig. 5.28 Block diagram of fitness function

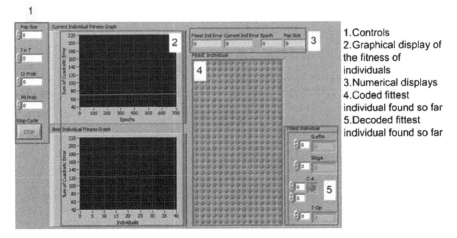

1.Controls
2.Graphical display of the fitness of individuals
3.Numerical displays
4.Coded fittest individual found so far
5.Decoded fittest individual found so far

Fig. 5.29 Front panel of the genetic programming example

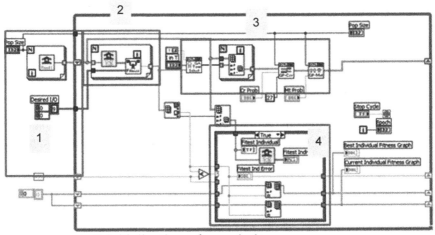

1. Initialization of population
2. Decodification and calculation of fitness
3. Selection, Crossover, Mutation
4. Selection of the fittest

Fig. 5.30 Block diagram of the genetic programming example

The fitness function (Figs. 5.28–5.30) compares a series of desired inputs and outputs with the corresponding performance of the controller, calculating the quadratic errors difference with each point, summing them and dividing by the number of evaluated points to obtain the fitness value for a given individual.

The *selection* function executes the tournament variation by randomly selecting a desired number of individuals and selecting the two fittest. This process is repeated until the same number of initial individuals is obtained.

Table 5.5 Controlled variables in the genetic programming example

Variable	Description
Pop Size	The number of individuals in the algorithm.
I in T	The number of individuals randomly taken for the tournament selection.
Cr Prob	The probability of crossing [0, 1].
Mt Prob	The probability of mutation for each bit [0, 1].

The *crossover* executes a one-stage interchange of tails, by taking two individuals from the mating pool, and depending on the possibility of crossing, the two individuals will or will not perform the crossover again. This process is repeated until the same number of initial individuals is obtained.

The *mutation* process executes a bit-to-bit operation on every one of the rules of each individual, and depending on the probability of mutation the bit will or will not change. During the execution of this algorithm, the individual with the best fitness is always stored to ensure that this information is never lost. Table 5.5 shows the variables to be controlled.

References

1. Wang F, et al. (2006) Design optimization of industrial motor drive power stage using genetic algorithms. Proceedings of CES/IEEE 5th I nternational Power Electronics and Motion Control Conference (IPEMC), 1–5 Aug 2006, vol 1, pp 14–16
2. Cho D-H, et al. (2001) Induction motor design of electric vehicle using a niching genetic algorithm. IEEE Trans Ind Appl USA 37(4):994–999
3. Cavalieri S (1999) A genetic algorithm for job-shop scheduling in a semiconductor manufacturing system. Proceedings of the 25th Annual Conference of the IEEE on Industrial Electronics Society (IECON) 1999, Italy, 29 Nov to 3 Dec 1999, vol 2, pp 957–961
4. Colla V, et al. (1998) Model parameter optimization for an industrial application: a comparison between traditional and genetic algorithms. Proceedings of the IEEE 2nd UKSIM European Symposium on Computer Modeling and Simulation, 8–10 Sept 2008, pp 34–39
5. Haupt RL, Haupt SE (1998) Practical genetic algorithms. Wiley-Interscience, New York
6. Mitchell M (1998) An introduction to genetic algorithms. MIT Press, Cambridge, MA
7. Cramer NL (1985) A representation for the adaptive generation of simple sequential programs. In: Grefenstette JJ (ed) Proceedings of the First International Conference on Genetic Algorithms and Their Applications. Erlbaum, Mahwah, NJ
8. Koza JR (1992) Genetic programming: on the programming of computers by means of natural selection. MIT Press, Cambridge, MA
9. Dolinsky JU, et al. (2007) Application of genetic programming to the calibrating of industrial robots. ScienceDirect Comput Ind 58(3):255–264
10. Langdon WB, Buxton BF (2003) The application of genetic programming for drug discovery in the pharmaceutical industry. EPSRC RIAS project with GlaxoSmithKline. London, UK, September 2003

Futher Reading

Dumitrescu D, et al. (2000) Evolutionary computation. CRC, Boca Raton, FL

Ghanea-Hercock R (2003) Applied evolutionary algorithms in Java. Springer, Berlin Heidelberg New York

Nedjah N, et al. (2006) Genetic systems programming theory and experiences. Springer, Berlin Heidelberg New York

Reeves CR, Rowe JE (2004) Genetic algorithms principles and perspectives: A guide to GA theory. Kluwer, Dordrecht

Chapter 6
Simulated Annealing, FCM, Partition Coefficients and Tabu Search

6.1 Introduction

In 1945 the construction of the first computer caused a revolution in the world. It was aimed to modify the interactions between Russia and the West. In the academic and research fields it brought back a mathematical technique known as statistical sampling, now referred to as the Monte Carlo method. S. Frankel and N. Metropolis created a model of a thermonuclear reaction for the Electronic Numerical Integrator and Computer (ENIAC), persuaded by the curiosity and interest of John von Neumann, a prominent scientist in that field.

The results of the model where obtained after the end of the World War I, and among the reviewers was Stan Ulam, who had an extensive background in mathematics and the use of statistical methods. He knew these techniques were no longer in use because of the length and tediousness of calculations. His research interest included pattern development in 2D games played with very simple rules. These techniques are now used in various industrial applications known as cellular automata.

Ulam and Neumann sent a proposal of the Monte Carlo method to the theoretical division leader of the Los Alamos Laboratory in New Mexico in 1947, which included a detailed outline of a possible statistical approach to solve the problem of neutron diffusion in fissionable material. The basic idea of the method was to generate a genealogical history of different variables in a process until a statistically valid picture of each variable was created.

The next step was to generate random numbers; here Neumann proposed a method called the *middle-square digits*. Once the random numbers are generated, they must be transformed into a non-uniform distribution desired for the property of interest. Solving problems using this method is easier than other approaches like differential equations, because one needs only to mirror the probability distribution into the search space of the problem at hand.

In 1947 the ENIAC was moved to the Ballistic Research Laboratory in Maryland, its permanent home. After the movement, there was an explosion in the use of the

P. Ponce-Cruz, F. D. Ramirez-Figueroa, *Intelligent Control Systems with LabVIEW™*   155
© Springer 2010

Monte Carlo method with this computer. The applications solved several questions of different branches of physics, and by 1949 there was a special symposium held on the method.

The Monte Carlo method gave birth to modern computational optimization problems. We can now see, as a natural consequence of electronic computers, the quick evolution of experimental mathematics, with the Monte Carlo method key to this achievement. It was at this point that mathematics achieved the twofold aspect of experiment and theory, which all other sciences enjoy.

As an example of the method, we can imagine a coconut shy. We want to know the probability of taking 10 shots at the coconut shy and obtain an even number of hits. The only information that we know is that there is a 0.2 probability of having a hit with a single shot. Using the Monte Carlo method we can perform a large number of simulations of taking 10 shots at the coconut shy. Next we can count the simulations with even number of hits and divide that number over the total number of simulations. By doing this we will get an approximation of the probability that we are looking for.

6.1.1 Introduction to Simulated Annealing

A combinatorial optimization problem strives to find the best or optimal solution, among a finite or infinite number of solutions. A wide variety of combinatorial problems have emerged from different areas such as physical sciences, computer science, and engineering, among others. Considerable effort has been devoted to construct and research methods for solving the performance of the techniques. Integer, linear, and non-linear programming have been the major breakthroughs in recent times.

Over the years it has been shown that many theoretical and practical problems belong to the class of *NP-complete problems*. A large number of these problems are still unsolved; there are two main options for solving them. On the one hand, if we strive for optimality the computation time will be very large; these methods are called *optimization methods*. On the other hand, we can search quick solutions with suboptimal performance, called *heuristic algorithms*. However, the difference between these methods is not very strict, because some types of algorithm can be used for both purposes.

Another way to classify algorithms is between general and tailored. While *general algorithms* are applicable to a wide range of problems, *tailored algorithms* use problem-specific information, restricting their applicability. The intrinsic problem is that for the former ones, for each type of combinatorial optimization problem, a new algorithm must be constructed.

6.1.2 Pattern Recognition

Recognizing and classifying patterns is a fundamental characteristic of human intelligence. It plays a key role in human perception as well as other levels of cognition. The field of study has evolved since the 1950s. *Pattern recognition* can be defined as a process by which we search for structures in data and classify them into categories such that the degree of association is high among structures of the same kind. Prototypical categories are usually characterized from past experience, and can be done by more than one structure.

Classification of objects falls in the category of *cluster analysis*, which plays a key role in pattern recognition. Cluster analysis it is not restricted to only pattern recognition, but is applicable to the taxonomies in biology and other areas, classification of information, and social groupings.

Fuzzy set theory has been used in pattern recognition since the mid-1960s. We can find three fundamental problems in pattern recognition, where most categories have vague boundaries. In general, objects are represented by a vector of measured values of r variables: $a = [a_1 \ldots a_r]$.

This vector is called a *pattern vector*, where a_i (for each $i \in N_r$) is a particular characteristic of interest. The first problem is concerned with representation of input data, which is obtained from the objects to be recognized, known as *sensing problems*. The second problem concerns the extraction of different features from the input data, in terms of the dimension of the pattern vector; they are called *feature extraction problems*. These features should characterize attributes, which determine the pattern classes.

The third problem involves the determination of optimal decision procedures for the classification of given patterns. Most of the time this is done by defining an *appropriate discrimination function* of patterns by assigning a real number to a pattern vector. Then, individual pattern vectors are evaluated by discrimination functions, and the classification is designed by the resulting number.

6.1.3 Introduction to Tabu Search

There are many problems that need to be solved by optimization procedures. Genetic algorithms (GA) or simulated annealing is used for that purpose. Tabu search (TS) is among the methods found in this field of optimization solutions. As its name suggests, tabu search is an algorithm performing the search in a region for the minimum or maximum solution of a given problem.

Searching is quite complicated because it uses a lot of memory and spends too much time in the process. For this reason, tabu search is implemented as an intelligent algorithm to take advantage of memory and to search more efficiently.

6.1.4 Industrial Applications of Simulated Annealing

We will briefly describe some industrial applications of the simulated annealing. Scheduling is always a difficult problem in the industry. Processes and logistics must be carefully combined to harmonize and increase production of plants. In 1997 A.P. Reynolds [1] and others presented a paper on simulated annealing for industrial applications. They optimized the scheduling process in order to optimize the resources of a manufacturing plant to meet the demand of different products.

S. Saika and other researchers [2] from Matsushita at the Advanced LSI Technology Development Center introduced a high-performance simulated annealing application to transistor placement. Called widely stepping simulated annealing, they applied it to the 1D transistor placement optimizations used in several industrial cells. They claim to have solutions as good as the standard algorithm and better, with a processing time one-thirtieth that of the normal simulated annealing.

R.N. Bailey, K.M. Garner, and M.F. Hobbs published a paper [3] showing the application of simulated annealing and GAs to solve staff scheduling problems. They use the algorithms to solve the scheduling of the work of staff with different skill levels, which is difficult to achieve because there is a large number of solutions. The results show that both simulated annealing and GAs can produce optimal and near-optimal solutions in a relatively short time for the nurse scheduling problem.

6.1.5 Industrial Applications of Fuzzy Clustering

Manufacturing firms have increased the use of industrial robots over the years. There has also been an increase in the number of robot manufacturers, offering a wide range of products. This is how M. Khouja and D.E. Booth [4] used a fuzzy clustering technique for the evaluation and selection of industrial robots given a specific application. They take into consideration real-world data instead to create the model.

B. Moshiri and S. Chaychi [5] use fuzzy logic and fuzzy clustering to model complex systems and identify non-linear industrial processes. They claim that their proposed advantage is simple, flexible and of high accuracy, easy to use and automatic. They applied this system to a heat exchanger.

6.1.6 Industrial Applications of Tabu Search

Tabu search has been widely used to optimize several industrial applications. For example, L. Zhang [6] and his team proposed a tabu search scheme to optimize the vehicle routing problem, with the objective of finding a schedule that will guarantee the safety of all vehicles. Their algorithm proved to be good enough compared with other more mature algorithms specially designed for the vehicle problem.

Artificial neural networks (ANNs) based on the tabu search algorithm have also been used by H. Shuang [7] to create a wind speed prediction model. A backpropagation neural network has its weights optimized using tabu search. Then the neural network is used as a model to predict the wind speed 1 hour ahead. It improved the prediction compared with a simple backpropagation neural network.

In 2007 J. Brigitte and S. Sebbah presented a paper [8] in which 3G networks are optimized. The location of primary bases and the core network link capacity is optimized. The dimensioning problem is modeled as a mixed-integer program and solved by a tabu search algorithm; the search criteria includes the signal-to-noise plus interference ratio. Primary bases are randomly located and after a few iterations their location is changed and the dimensioning optimized.

This base optimization problem was previously addressed by C.Y. Lee and published in a paper in 2000 [9]. He also aimed to minimize the number of base stations used and its location in an area covered by cellular communications. The results presented show that a 10 % in cost reduction is achieved, and between a 10 and 20% of cost reduction in problems with 2500 traffic demand areas with code division multiple access (CDMA) systems.

6.2 Simulated Annealing

It was in 1982 and 1983 that Kirkpatrick, Gelatt and Vecchi introduced the concepts of annealing in combinatorial optimization. It was also independently presented in 1985 by Černy. The concepts are based on the physical annealing process of solids and the problem of solving large optimization problems.

Annealing is a physical process where a substance is heated and cooled in a controlled manner. The results obtained by this process are strong crystalline structures, compared to structures obtained by fast untempered cooling, which result in brittle and defective structures. For the optimization process the structure is our encoded solution, and the temperature is used to determine how and when new solutions are accepted. The process contains two steps [4, 10]:

1. *Increase* temperature of the heat bath to a maximum value at which the solid melts.
2. *Carefully* decrease the temperature of the heat bath until particles arrange themselves in the ground state of the solid.

When the structure is in the liquid phase all the particles of the solid arrange themselves in a random way. In the ground state the particles are arranged in a highly structured lattice, leaving the energy of the system at its minimum. This ground state of the solid is only obtained if the maximum temperature is sufficiently high and the cooling is sufficiently low, otherwise, the solid will be frozen into a metastable state rather than the ground state.

Computer simulation methods from condensed matter physics are used to model the physical annealing process. Metropolis and others introduced a simple algorithm to simulate the evolution of a solid in a heat bath at thermal equilibrium. This algorithm is based on Monte Carlo techniques, which generate a sequence of states of the solid.

These states act as the following: given the actual state i of the solid that has energy E_i, the subsequent state j is generated by applying a perturbation mechanism which transforms the present state into the next state causing a small distortion, like displacing a particle. For the next state E_j, if the *energy difference* $E_j - E_i$ is less than or equal to zero, then the j is accepted as the current state. If the energy difference is greater than zero, then the j state is accepted with a certain probability, given by: $e^{\left(\frac{E_i - E_j}{k_B T}\right)}$.

Here, T denotes the temperature of the heat bath, and k_B is a constant known as the Boltzmann constant. We will now describe the *Metropolis criterion* used as the acceptance rule. The algorithm that goes with it is known as the *Metropolis algorithm*.

If the temperature is lowered sufficiently slowly, then the solid will reach thermal equilibrium at each temperature. In the Metropolis algorithm this is achieved by generating a large number of transitions at a given temperature value. The thermal equilibrium is characterized by a Boltzmann distribution, which gives the probability of the solid being in the state i with an energy E_i at temperature T:

$$P_T \{X = i\} = \frac{1}{Z(T)} e^{\left(-\frac{E_i}{k_B T}\right)}, \tag{6.1}$$

where X is a stochastic variable that denotes the state of the solid in its current form, and $Z(T)$ is a *partition function*, defined by:

$$Z(T) = \sum_j e^{\left(-\frac{E_j}{k_B T}\right)}. \tag{6.2}$$

The sum will extend over all the possible states. The simulated annealing algorithm is very simple and can be defined in six steps [11], as shown in Fig. 6.1.

1. *Initial Solution*
   The initial solution will be mostly a random one and gives the algorithm a base from which to search for a more optimal solution.
2. *Assess Solution*
   Consists of decoding the current solution and performing whatever action is necessary to evaluate it against the given problem.
3. *Randomly Tweak Solution*
   Randomly modify the working solution, which depends upon the encoding.
4. *Acceptance Criteria*
   The working solution is compared to the current solution, if the working one has less energy than the current solution (a better solution) then the working solution is copied to the current solution and the temperature is reduced. If the working

Fig. 6.1 Simulated annealing algorithm

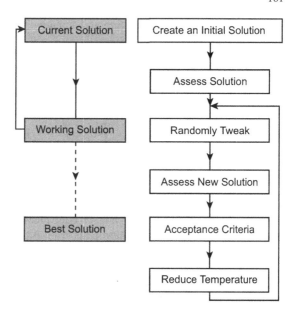

solution is worse than the current one, the acceptance criteria is evaluated to determine what to do with the current solution. The probability is based on (6.3):

$$P(\delta E) = e^{-\frac{\delta E}{T}}, \tag{6.3}$$

which means that at higher temperatures poorer solutions are accepted in order to search in a wider range of solutions.

5. *Reduce Temperature*

   After a certain number of iterations the temperature is decreased. The simplest way is by means of a geometric function $T_{i+1} = \alpha T_i$, where the constant α is less than one.

6. *Repeat*

   A number of operations are repeated at a single temperature. When that set is reduced the temperature is reduced and the process continues until the temperature reaches zero.

6.2.1 Simulated Annealing Algorithm

We need to assume an analogy between the physical system and a combinatorial optimization problem, based on the following equivalences:

- Solutions in a combinatorial optimization problem are equivalent to states of a physical system.
- The energy of a state is the cost of a solution.

The *control parameter* is the temperature, and with all these features the simulated annealing algorithm can now be viewed as an iteration of the Metropolis algorithm evaluated at decreasing values of the control parameters. We will assume the existence of a neighborhood structure and a generation mechanism; some definitions will be introduced.

We will denote an instance of a combinatorial optimization problem by (S, f), and i and j as two solutions with their respective costs $f(i)$ and $f(j)$. Thus, the acceptance criterion will determine if j is accepted by i by applying the following *acceptance probability*:

$$P_c \text{ (accepted } j) = \begin{cases} 1 & \text{if } f(j) \leq f(i) \\ e^{\left(\frac{f(i)-f(j)}{c}\right)} & \text{if } f(j) > f(i) \end{cases} . \quad (6.4)$$

where here $c \in R^+$ denotes the control parameter. The generation mechanism corresponds to the perturbation mechanism equivalent at the Metropolis algorithm and the acceptance criterion is the Metropolis criterion.

Another definition to be introduced is the one of *transition*, which is a combined action resulting in the transformation of a current solution into a subsequent one. For this action we have to follow the next two steps: (1) application of the generation mechanism, and (2) application of the acceptance criterion.

We will denote c_k as the value of the control parameter and L_k as the number of transitions generated at the kth iteration of the Metropolis algorithm. A formal version of the simulated annealing algorithm [5] can be written in pseudo code as shown in Algorithm 6.1.

Algorithm 6.1

```
SIMULATED ANNEALING
init:
k = 0
i = i_start
repeat
  for l = 1 to L_k do
  GENERATE j from S_i :
  if f (j) ≤ f (i) then i = j
    else
      if e^((f(i)−f(j))/c_k) > rand [0, 1) then i = j
      k = k + 1
      CALCULATE LENGTH (L_k)
      CALCULATE CONTROL (L_k)
  until stopcriterion
end
```

The probability of accepting perturbations is implemented by comparing the value of $e^{f(i)-f(j)}/c$ with random numbers generated in $(0, 1)$. It should also be obvious that the speed of convergence is determined by the parameters L_k and c_k.

A feature of simulated annealing is that apart from accepting improvements in cost, it also accepts, to a limited extent, deteriorations in cost. With large values of c large deteriorations or changes will be accepted. As the value of c decreases, only smaller deteriorations will be accepted. Finally, as the value approaches zero, no perturbations will be accepted at all. This means that the simulated annealing algorithm can escape from local minima, while it still is simple and applicable.

6.2.2 Sample Iteration Example

Let us say that the current environment temperature is 50 and the current solution has an energy of 10. The current solution is perturbed, and after calculating the energy the new solution has an energy of 20. In this case the energy is larger, thus worse, and we must therefore use the acceptance criteria. The delta energy of this sample is 10. Calculating the probability we will have:

$$P = e^{\left(-\frac{10}{50}\right)} = 0.818731 . \tag{6.5}$$

So for this solution it will be very probable that the less ideal solution will be propagated forward. Now taking our schedule at the end of the cycle, the temperature will be now 2 and the energies of 3 for the current solution, and 7 for the working one. The delta energy of the sample is 4. Therefore, the probability will be:

$$P = e^{\left(-\frac{4}{2}\right)} = 0.135335 . \tag{6.6}$$

In this case, it is very unlikely that the working solution will be propagated in the subsequent iterations.

6.2.3 Example of Simulated Annealing Using the Intelligent Control Toolkit for LabVIEW

We will try to solve the N-queens problem (NQP) [3], which is defined as the placement of N queens on an $N \times N$ board such that no queen threatens another queen using the standard chess rules. It will be solved in a 30×30 board.
Encoding the solution. Since each column contains only one queen, an N-element array will be used to represent the solution.
Energy. The energy of the solution is defined as the number of conflicts that arise, given the encoding. The goal is to find an encoding with zero energy or no conflicts on the board.
Temperature schedule. The temperature will start at 30 and will be slowly decreased with a coefficient of 0.99. At each temperature change 100 steps will be performed.

Fig. 6.2 Simulated annealing VIs

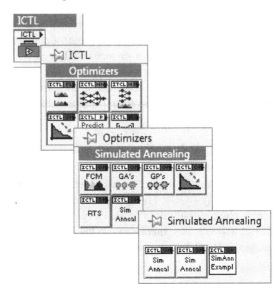

The initial values are: initial temperature of 30, final temperature of 0.5, alpha of 0.99, and steps per change equal to 100. The VIs for the simulated annealing are found at: Optimizers ≫ Simulated Annealing, as shown in Fig. 6.2.

The front panel is like the one shown in Fig. 6.3. We can choose the size of the board with the *MAX_LENGTH* constant. Once a solution is found the green LED *Solution* will turn on. The initial constants that are key for the process are introduced in the cluster *Constants*. We will display the queens in a 2D array of bits. The *Current*, *Working* and *Best* solutions have their own indicators contained in clusters.

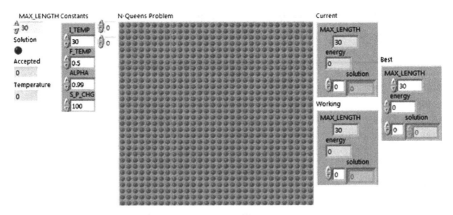

Fig. 6.3 Front panel for the simulated annealing example

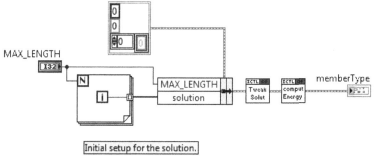

Fig. 6.4 Block diagram for the generation of the initial solution

Our initial solution can be created very simply; each queen is initialized occupying the same row as its column. Then for each queen the column will be varied randomly. The solution will be tweaked and the energy computed. Figure 6.4 shows the block diagram of this process.

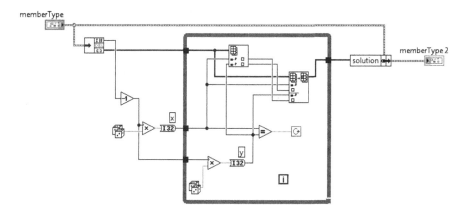

Fig. 6.5 Code for the tweaking process of the solution

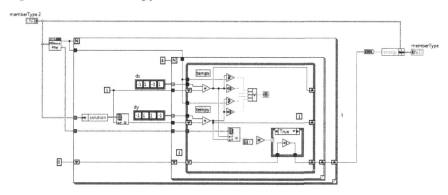

Fig. 6.6 Code for the computation of energy

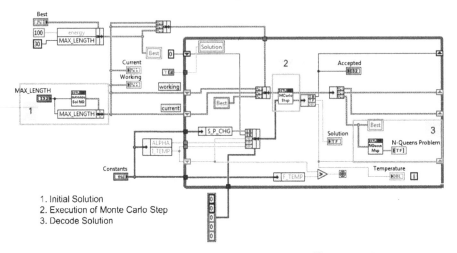

1. Initial Solution
2. Execution of Monte Carlo Step
3. Decode Solution

Fig. 6.7 Block diagram of the simulated annealing example for the N-queen problem

The tweaking is done by the code shown in Fig. 6.5; basically it randomizes the position of the queens. The energy is computed with the following code. It will try to find any conflict in the solution and assess it. It will select each queen on the board, and then on each of the four diagonals looking for conflicts, which are other queens in the path. Each time one is found the conflict variable is increased. In Fig. 6.6 we can see the block diagram. The final block diagram is shown in Fig. 6.7.

6.3 Fuzzy Clustering Means

In the field of optimization, fuzzy logic has many beneficial properties. In this case, *fuzzy clustering means* (FCM), known also as *fuzzy c-means* or *fuzzy k-means*, is a method used to find an optimal clustering of data.

Suppose, we have some collection of data $\mathbf{X} = \{x_1, \ldots, x_n\}$, where every element is a vector point in the form of $x_i = (x_i^1, \ldots, x_i^p) \in \mathfrak{R}^p$. However, data is spread in the space and we are not able to find a clustering. Then, the purpose of FCM is to find clusters represented by their own centers, in which each center has a maximum separation from the others. Actually, every element that is referred to as clusters must have the minimum distance between the cluster center and itself. Figure 6.8 shows the representation of data and the FCM action.

At first, we have to make a partition of the input data into c subsets written as $P(X) = \{U_1, \ldots, U_c\}$, where c is the number of partitions or the number of clusters that we need. The partition is supposed to have fuzzy subsets U_i. These subsets must satisfy the conditions in (6.7) to (6.8):

$$\sum_{i=1}^{c} U_i(x_k) = 1, \quad \forall x_k \in X \tag{6.7}$$

$$0 < \sum_{k=1}^{n} U_i(x_k) < n. \tag{6.8}$$

The first condition says that any element x_k has a fuzzy value to every subset. Then, the sum of membership values in each subset must be equal to one. This condition suggests to elements that it has some membership relation to all clusters, no matter how far away the element to any cluster. The second condition implies that every cluster must have at least one element and every cluster cannot contain all elements in the data collection. This condition is essential because on the one hand, if there are no elements in a cluster, then the cluster vanishes.

On the other hand, if one cluster has all the elements, then this clustering is trivial because it represents all the data collection. Thus, the number of clusters that FCM can return is $c = [2, n - 1]$. FCM need to find the centers of the fuzzy clusters. Let $v_i \in \mathfrak{R}^p$ be the vector point representing the center of the ith cluster, then

$$v_i = \frac{\sum\limits_{k=1}^{n} [U_i(x_k)]^m x_k}{\sum\limits_{k=1}^{n} [U_i(x_k)]^m}, \quad \forall i = 1, \dots, c, \tag{6.9}$$

where $m > 1$ is the *fuzzy parameter* that influences the grade of the membership in each fuzzy set. If we look at (6.9), we can see that it is the weighted average of the data in U_i. This expression tells us that centers may or may not be any point in the data collection.

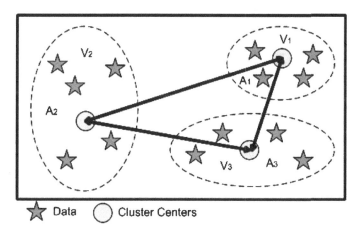

Fig. 6.8 Representation of the FCM algorithm

Actually, FCM is a recursive algorithm, and therefore needs an objective function that estimates the optimization process. We may say that the objective function $J_m(P)$ with grade m of the partition $P(X)$ is shown in (6.10):

$$J_m(P) = \sum_{k=1}^{n} \sum_{i=1}^{c} [U_i(x_k)]^m \|x_k - v_i\|^2. \tag{6.10}$$

This objective function represents a measure of how far the centers are from each other, and how close the elements in each center are. For instance, the smaller the value of $J_m(P)$, the better the partition $P(X)$. In these terms, the goal of FCM is to minimize the objective function.

We present the FCM algorithm developed by J. Bezdek for solving the clustering data. At first, we have to select a value $c = [2, n - 1]$ knowing the data collection X. Then, we have to select the fuzzy parameter $m = (1, \infty)$. In the initial step, we select a partition $P(X)$ randomly and propose that $J_m(P) \to \infty$. Then, the algorithm calculates all cluster centers by (6.9). Then, it updates the partition by the following procedure: for each $x_k \in X$ calculate

$$U_i(x_k) = \left[\sum_{j=1}^{c} \left(\frac{\|x_k - v_i\|^2}{\|x_k - v_j\|^2} \right)^{\frac{1}{m-1}} \right]^{-1}, \quad \forall i = 1, \ldots, c. \tag{6.11}$$

Finally, the algorithm derives the objective function with values found by (6.9) and (6.11), and it is compared with the previous objective function. If the difference between the last and current objective functions is close to zero (we say $\varepsilon \geq 0$ is a small number called the *stop criterion*), then the algorithm stops. In another case, the algorithm recalculates cluster centers and so on. Algorithm 6.2 reviews this discussion. Here $n = [2, \infty)$, $m = [1, \infty)$, U are matrixes with the membership functions from every sample of the data set to each cluster center. P are the partition functions.

Algorithm 6.2	FCM procedure
Step 1	Initialize time $t = 0$.
	Select numbers $c = [2, n - 1]$ and $m = (1, \infty)$.
	Initialize the partition $P(X) = \{U_1, \ldots, U_c\}$ randomly.
	Set $J_m(P)^{(0)} \to \infty$.
Step 2	Determine cluster centers by (6.9) and $P(X)$.
Step 3	Update the partition by (6.11).
Step 4	Calculate the objective function $J_m(P)^{(t+1)}$.
Step 5	If $J_m(P)^{(t)} - J_m(P)^{(t+1)} > \varepsilon$ then update $t = t + 1$ and go to *Step 2*.
	Else, *STOP*.

Example 6.1. For the data collection shown in Table 6.1 with 20 samples. Cluster in three subsets with a FCM algorithm taking $m = 2$.

Table 6.1 Data used in Example 6.1

Number	X data	Number	X data	Number	X data	Number	X data
1	255	6	64	11	58	16	80
2	67	7	64	12	96	17	80
3	67	8	71	13	96	18	71
4	74	9	71	14	87	19	71
5	74	10	58	15	87	20	62

Fig. 6.9 Block diagram of the initialization process

Fig. 6.10 Block diagram of partial FCM algorithm

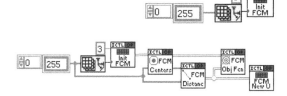

Solution. The FCM algorithm is implemented in LabVIEW in several steps. First, following the path ICTL ≫ Optimizers ≫ FCM ≫ FCM methods ≫ **init_fcm.vi**. This VI initializes the partition. In particular, it needs the number of clusters (for this example 3) and the size of the data (20). The output pin is the partition in matrix form. Figure 6.9 shows the block diagram. The 1D array is the vector in which the twenty elements are located.

Then, we need to calculate the cluster centers using the VI at the path ICTL ≫ Optimizers ≫ FCM ≫ FCM methods ≫ **centros_fcm.vi**. One of the input pins is the matrix **U** and the other is the data. The output connections are referred to as U^2 and the cluster centers *Centers*. Then, we have to calculate the objective function. The VI is in ICTL ≫ Optimizers ≫ FCM ≫ FCM methods ≫ **fun_obj_fcm.vi**. This VI needs two inputs, the U^2 and the distances between elements and centers. The last procedure is performed by the VI found in the path ICTL ≫ Optimizers ≫ FCM ≫ FCM methods ≫ **dist_fcm.vi**. It needs the cluster centers and the data. Thus, **fun_obj_fcm.vi** can calculate the objective function with the distance and the partition matrix powered by two coming from the previous two VIs. In the same way, the partition matrix must be updated by the VI at the path ICTL ≫ Optimizers ≫ FCM ≫ FCM methods ≫ **new_U_fcm.vi**. It only needs the distance between elements and cluster centers. Figure 6.10 shows the block diagram of the algorithm.

Of course, the recursive procedure can be implemented with either a *while-loop* or a *for-loop* cycle. Figure 6.11 represents the recursive algorithm. In Fig. 6.11 we create a *Max Iterations* control for number of maximum iterations that FCM could reach. The *Error* indicator is used to look over the evaluation of the objective function and *FCM Clusters* represents graphically the fuzzy sets of the partition matrix found. We see at the bottom of the while-loop, the comparison between the last error and the current one evaluated by the objective function. □

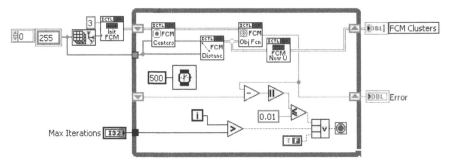

Fig. 6.11 Block diagram of the complete FCM algorithm

6.4 FCM Example

This example will use previously gathered data and classify it with the FCM algorithm; then we will use T-ANNs to approximate each cluster. The front panel is
shown in Fig. 6.12.

We will display the normal FCM clusters in a graph, the approximated clusters in
another graph and the error obtained by the algorithm in an *XY* graph. We also need
to feed the program with the number of neurons to be used for the approximation,
the number of clusters and the maximum allowed iterations. Other information can
be displayed like the centers of the generated clusters and the error between the ap-

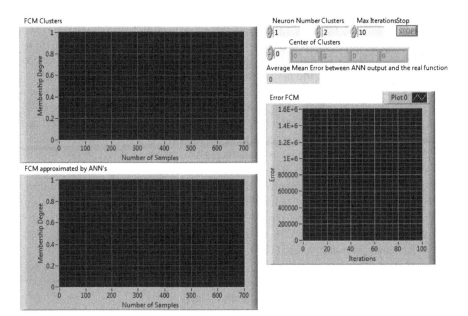

Fig. 6.12 Front panel of the FCM example

proximated version of the clusters and the normal one. This example can be located at Optimizers ≫ FCM ≫ **Example_FCM.vi** where the block diagram can be fully inspected, as seen in Fig. 6.13 (with the results shown in Fig. 6.14).

This program takes information previously gathered, then initializes and executes the FCM algorithm. It then orders the obtained clusters and trains a T-ANNs with

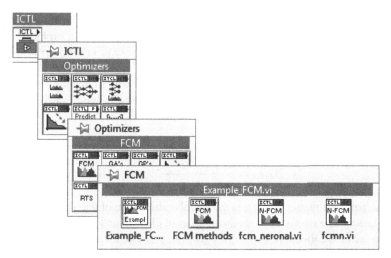

Fig. 6.13 VIs for the FCM technique

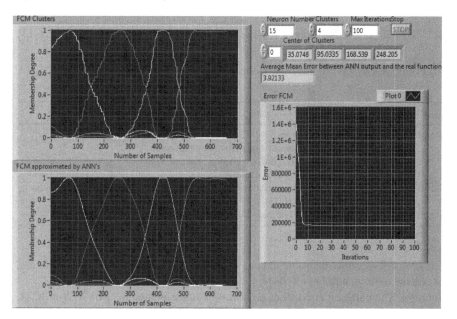

Fig. 6.14 The FCM program in execution

the information of each cluster. After that the T-ANNs are evaluated, and the average mean error between the approximated and the real clusters are calculated.

6.5 Partition Coefficients

FCM described in Algorithm 6.1 is very useful in pattern recognition techniques. However, no matter which application is being developed, FCM has a problem: what could be the value for the number of clusters? The answer is the *partition coefficient*.

Partition coefficient (PC) is a method used to validate how well a clustering algorithm has identified the structure presented in the data, and how it represents it into clusters. This small algorithm is based on the following:

$$PC\,(\mathbf{U};c) = \frac{\sum_{j=1}^{n}\sum_{i=1}^{c}\left(u_{ij}\right)^{2}}{n}, \tag{6.12}$$

where $\mathbf{U}$ is the partition matrix and u_{ij} is the membership value of the jth element of the data related to the ith cluster, c is the number of clusters and n is the number of elements in the data collection. From this equation, it can be noted that the closer the PC is to 1, the better classified the data is considered to be. The optimal number of clusters can be denoted at each c by Ω_c using (6.13):

$$\max_{c}\left[\max_{\Omega_c \in U}\{PC\,(U;c)\}\right]. \tag{6.13}$$

Algorithm 6.3 shows the above procedure.

Algorithm 6.3	Partition coefficient
Step 1	Initialize $c = 2$. Run FCM or any other clustering algorithm.
Step 2	Calculate the partition coefficient by (6.12).
Step 3	Update the value of clusters $c = c + 1$.
Step 4	Run until no variations at PC are found and obtain the optimal value of clusters by (6.13).
Step 5	Return the optimal value c and *STOP*.

Example 6.2. Assume the same data as in Example 6.1. Run the PC algorithm and obtain the optimal number of clusters.

Solution. The partition coefficient algorithm is implemented in LabVIEW at ICTL ≫ Optimizers ≫ Partition Coeff. ≫ **PartitionCoefficients.vi**. On the inside of this VI, the FCM algorithm is implemented. So, the only thing we have to do is to connect the array of data and the number of clusters at the current iteration. Figure 6.15

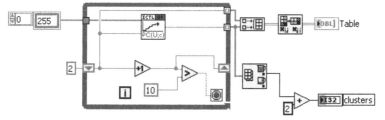

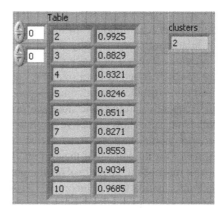

Fig. 6.15 Block diagram finding the optimal number of clusters

Fig. 6.16 Front panel of
Example 6.2 showing the
optimal value for clusters

is the block diagram of the complete solution of this example. In this way, we initialize the number of clusters in 2 and at each iteration, this number is increased.
The number 10 is just for stopping the process when the number of clusters is larger
than this. Finally, in *Table* we find the evaluated PC at each number of clusters and
clusters indicates the number of optimal clusters for this particular data collection.
Figure 6.16 shows the front panel of this example. The solution for this data collection is 2 clusters.   □

6.6 Reactive Tabu Search

6.6.1 Introduction to Reactive Tabu Search

The word *tabu* means that something is dangerous, and taking it into account involves a risk. This is not used to avoid certain circumstances, but instead is used in
order to prohibit features, for example, until the circumstances change. As a result,
tabu search is the implementation of intelligent decisions or the responsive exploration in the search space.

The two main properties of tabu search are *adaptive memory* and *responsive
exploration*. The first term refers to an adaptation of the memory. Not everything

is worth remembering, but not everything is worth forgetting either. This property is frequently used to make some subregion of the search space tabu. Responsive exploration is a mature decision in what the algorithm already knows, and can be used to find a better solution. The latter, is related to the rule by which tabu search is inspired: a bad strategic choice can offer more information than a good random choice. In other words, sometimes is better to make a choice that does not qualify as the best one at that time, but it can be used to gather more information than the better solution at this time.

More precisely, tabu search can be described as a method designed to search in not so feasible regions and is used to intensify the search in the neighborhood of some possible optimal location.

Tabu search uses memory structures that can operate in a distinct kind of region, which are recency, frequency, quality, and influence. The first and the second models are *how recent* and *at what frequency* one possible solution is performed. Thus, we need to record the data or some special characteristic of that data in order to count the frequency and the time since the same event last occurred. The third is *quality*, which measures how attractive a solution is. The measurement is performed by features or characteristics extracted from data already memorized. The last structure is *influence*, or the impact of the current choice compared to older choices, looking at how it reaches the goal or solves the problem. When we are dealing with the direct data information stored, memory is *explicit*. If we are storing characteristics of the data we may say that memory is *attributive*.

Of course, by the adaptive memory feature, tabu search has the possibility of storing relevant information during the procedure and forgetting the data that are not yet of interest. This adaptation is known as *short term memory* when data is located in memory for a few iterations; *long term memory* is when data is collected for a long period of time.

Other properties of tabu search are the *intensification* and *diversification* procedures. For example, if we have a large search region, the algorithm focuses on one possible solution and then the intensification procedure explores the vicinity of that solution in order to find a better one. If in the exploration no more solutions are optimally found, then the algorithm diversifies the solution. In other words, it leaves the vicinity currently explored and goes to another region in the search space. That is, tabu search explores large regions choosing small regions in certain moments.

6.6.2 *Memory*

Tabu search has two types of memory: short term and long term-based memories. In this section we will explain in more detail how these memories are used in the process of optimizing a given problem.

To understand this classification of memory, it is necessary to begin with a mathematical description. Suppose that the search space is V so $x \in V$ is an element of

Fig. 6.17 Search space of the
tabu search

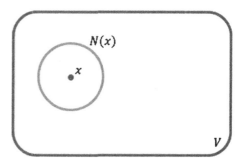

the solution in the search space. The algorithm searches in the vicinity of x known
as $N(x)$ where $N(x) \subset V$. Figure 6.17 shows these terms.

6.6.2.1 Short Term Memory

The first memory used in tabu search is the short term. When we are searching in
the vicinity $N(x)$, and we try to do it as fast as possible. Another method is the
so-called *steepest descent method* that can be explained as follows: we pick up an
element x, then we take an objective function $f(x)$ and store this value. In the
next iteration we look for an element $x' \in N(x)$ and evaluate the function $f(x')$.
If $f(x') < f(x)$ then x' is the new optimal solution. We repeat until the condition
is not true. Therefore, the solution of that method is known as the local optimum,
because the solution is the optimal one in the vicinity but not in the entire search
space.

This process is very expensive computationally. Therefore, a short term memory
is used. We try to delimit the search space by the vicinity of an element of that space.
Then, we try to minimize as much as possible the search space. How can we do it?
If some features of the solution are really known, then it is easy to deduce that some
solutions are prohibited. For example, the last element evaluated is prohibited from
being selected. This is a simple example, but some other characteristics might be
applied to avoid the selection of these possible solutions. So, there exists a subspace
of the vicinity $N(x)$ that is a tabu list, namely T. The new vicinity is characterized as
$N^*(x) = N(x)/T$. In this way, tabu search is an algorithm that explores a dynamic
vicinity of the solution.

The tabu list or the tabu space is stored in short term memory. One of the simplest
uses of this memory is in the recent process. If some solution x is evaluated then the
next few iterations are prohibited.

In the same manner, when we talk about iterations of selecting new elements
to evaluate, we are trying to say that the change or the *move* between the current
solution and the following is evaluated to know if this change is really useful or
not. The dynamic vicinity of movement is distinguished by two classifications: the
vicinity of permissible moves and the tabu moves. Then, by the attributes of the

elements x we can determine if some move can be added to the permissible vicinity $N^*(x)$ or if the move might be dropped to the T subspace.

In order to assign the time of prohibition of some special element, the tabu tenure is created. The tabu tenure is a value t in the interval $[t_{min}, t_{max}]$ that describes the number of prohibited iterations remaining until the element will be reused. This value can be assigned randomly or in a more systematic form.

Example 6.3. Let $V = \{9, 4, 6, 1, 8, 2\}$ be the values of the search space and the vicinity $N(x) = \{4, 8, 9\}$ with value $x = 6$, considering the vicinity with a radius of 3. Then, assume the tabu list of the entire domain as $T(V) = \{0, 0, 5, 0, 4, 0\}$. Suppose that $t \in [0, 5]$ and when some element is selected, the tabu tenure is 5 and all the other components of T are decreased by one. (a) What is the permissible vicinity $N^*(6)$? (b) What was the last value selected before $x = 6$? (c) If we are looking around $x = 4$ what could be the entire vicinity $N(4)$?

Solution. (a) Looking at the tabu set, we know that all 0 values mean that elements in that position are permitted. So, the possible elements that can be picked up are $N^*(V) = \{9, 4, 1, 2\}$. But, we are searching in the vicinity $N(6) = \{4, 8, 9\}$. Then, $N^*(6) = N^*(V) \cap N(6) = \{9, 4, 1, 2\} \cap \{4, 8, 9\}$. Finally, the permissible set around the element 6 is: $N^*(6) = \{4, 9\}$.

(b) As we can see, the current element is 6, then in the tabu list this element has a value of $t = 5$. This matches with the procedure defined in the example. Actually, in this way all other values were decreased by one. Therefore, if we are trying to look before the current value, the tabu list must be $T(V)_{before\_6} = T(V)_{current} + 1$. In other words, $T(V)_{before\_6} = \{d, d, 0, d, 5, d\}$, where d is a possible value different or equal to zero because we are uncertain. Actually, if 5 is the tenure assigned for the current selection, then the current selection before the 6-element is 8. This is the reason why in the $N^*(6)$, this element does not appear.

(c) Obviously, we just need to look around 4 with a radius of 3. So, the vicinity is $N(4) = \{1, 2, 6\}$.

With the quality property, some moves can be reinforced. For example, if some element is selected, then the tabu tenure is fired and the element will be in the tabu list. But, if the element has a high quality, then the element could be promoted to a permissible value. This property of the short term memory is then useful in the process of tabu search.   □

6.6.2.2 Long Term Memory

In the same way as the short term, this type of memory is used in order to obtain more attributes for the search procedure. However, long term memory is focused on storing information about the past history.

The first implementation is the frequency dimension of memory. In other words, we can address some data information (explicitly or attributively) and after some iterations or movements, try to analyze the frequency with which this data has appeared. In this way, there exists an array of elements named *transition measure*, which counts the number of iterations that some element was modified. Another array linked with the frequency domain is the *residence measure* and it stores the number of iterations that some element has taken part in the solution.

What is the main purpose of long term memory? It is easy to answer: frequency or what remains the same, transition and residence are measures of how attractive some element is to be included in the vicinity when the frequency is high. On the other hand, if the frequency is low, the element related to that measure has to be removed or placed in the tabu list. Then, long term memory is used to assign attributes to elements in order to discard or accept those in the dynamic vicinity.

Example 6.4. Let $V = \{9, 4, 6, 1, 8, 2\}$ be the values of the search space. Suppose that $tr = \{0, 2, 5, 2, 1, 4\}$ is the transition array and $r = \{3, 1, 0, 0, 2, 0\}$ is the residence array. (a) Have any of the elements in V ever been in the best solution thus far? Which one? (b) How many iterations was the element 8 taken into account in the best solution thus far? (c) Can you know at which iteration the best solution was found?

Solution. (a) Yes. There are three elements that have been in the best solution. These elements are 9, 4 and 8 because its residence elements are distinct from zero. In fact, element 9 has been in the solution three times, element 4 has been one time, and element 8 has been two times.

(b) By the transition array we know that element 1 associated to element 8 refers to the fact that this element has been in the best solution only one time. Element 6 has been five times in the best solution so far.

(c) Yes, we can know at which iteration the best solution was found. Suppose that the current iteration is t. The transition array updates if some element has been in the best solution, but if the best solution is modified by the best solution so far, obviously the transition array is initialized. Then, the maximum number of counts that the current transition array has is 5. So, the best solution so far was found at iteration $t_{\text{best}} = t - 5$. $\square$

6.6.2.3 Intensification and Diversification

In the introduction, we said that tabu search is a method that tries to find the best solution in short periods of time without searching exhaustively. These characteristics are offered in some way by the two types of memory described above and by the two following procedures.

The first one is *intensification*. Suppose that we have a search space V, which has n subspaces W_i, $\forall i = 1, \ldots, n$, and that these have no intersection and the two characteristics apply: $(W_i \cap W_j) = \emptyset, \forall i \neq j$ and $W_i \subset V$. So, $V = \{W_i \cup W_j\}, \forall i \neq j$. This means that the search space can be divided by n regions. We can think of one of these subspaces as in the vicinity of an element of V, so-called $W_i = N(x_i)$. We also know that tabu search looks in this vicinity in order to find the local optimum x_i'. Then, this local optimum is stored in the memory and the process is repeated in another vicinity $N(x_j)$. Suppose now that the algorithm recorded k local optimum elements $\{x_1', \ldots, x_j', \ldots, x_k'\}$. The intensification is thus the process in which the algorithm visits the vicinity of each of the local optimum elements recorded by means, and looks at the vicinity $N(x_i'), \forall i = 1, \ldots, k$.

In other words, the intensification procedure is a subroutine of the tabu search that tries to find a better solution in the vicinity of the best solutions (local optimum values) so far. This means that it intensifies the exploration.

The second procedure is *diversification*. Let us suppose that we have the same environment as in the intensification procedure. The question is how the algorithm explores distinct places in order to record local optimum values. Diversification is the answer. When some stop searching threshold is fired, the local optimum is recorded and then the algorithm accepts the option to search in a different region, because there may be some other solutions (either better or just as good as the local optimum found so far) placed in other regions.

To make this possible, the algorithm records local optimum values and evaluates permissible movements in the entire region. Therefore, the tabu list has the local optimum elements found so far and all the vicinities of these values. In this way, more local optimum elements mean fewer regions in which other solutions could be found. Or, in the same way, the set of permissible movements comes to be small.

As we can see, the intensification procedure explores regions in which good solutions were found and the diversification procedure permits the exploration into unknown regions. Thus, the algorithm searches as much as it can and restricts all movements when possible to use less time.

6.6.2.4 Tabu Search Algorithm

Tabu search has several modifications in order to get the best solution as fast as possible. In this way, we explain first the general methodology and then we explain in more detail the modification known as reactive tabu search.

6.6.2.5 Simple Tabu Search

First, we need a function that describes the solution. This function is constructed by elements. If the solution has n-dimension size, then the function must have n elements of the form $f = \{f_1, \ldots, f_n\}$. We refer to the function configuration at time t with $f^{(t)} = \{f_1^{(t)}, \ldots, f_n^{(t)}\}$. Then, we aggregate some terminology shown in Table 6.2.

We associate to each of the elements in $f^{(t)}$, a permissible move referred to as $\mu_i, \forall i = 1, \ldots, n$; they are well defined in a set of permissible moves A. All other moves are known as tabu elements displayed in the set τ. Actually, the complement of A is τ.

First, all movements are permitted, so A is all the search space and $\tau = \emptyset$. A configuration is selected randomly. In this case, we need a criterion in order to either intensify the search or to diversify it. The criterion selected here is to know if the current configuration has been selected before. If the frequency of this configura-

Table 6.2 Terminology for simple tabu search

Symbol	Description
t	Iteration counter
f	Configuration at time t
μ	Set of elementary moves
L	Last iteration when the move was applied
Π	Last iteration when the configuration was used
Φ	Number of times the configuration has been visited during the search
f_b	Best configuration known
E_b	Best energy known
A	Set of admissible moves
C	Chaotic moves
S	Subset of A
τ	Set of tabu moves, non-admissible moves
T	Prohibition period

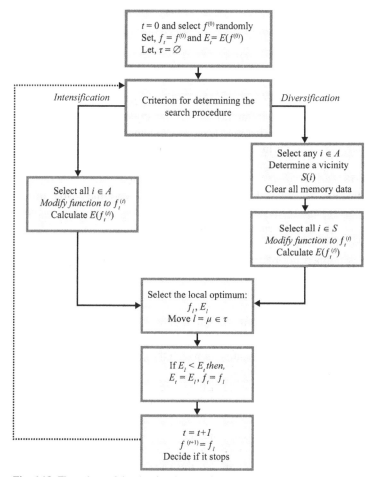

Fig. 6.18 Flow chart of the simple tabu search

tion is high, then the algorithm needs to diversify it; otherwise, the intensification procedure is defined as the following.

Using the current configuration, we need to evaluate all possible elements. An energy function $E(f_i^{(t)})$, $\forall i = 1, \ldots, |S|$ is evaluated for each of the configurations, done by selecting all the possible moves ($|S|$ means the cardinality or the number of elements in the subset of permissible moves $S \subset A$). The criterion is to select the configuration that minimizes the energy function. After that, this configuration is labeled as the best configuration f_b and the energy is stored in a variable named the best energy known so far E_b.

If we need a diversification procedure, the algorithm erases all memory data and makes a searching process in a distinct region of the search space, which tries to find a local optimum f_b. The best energy and the best configuration is then actualized with this process.

Finally, the iteration time is incremented and the procedure is done until some stop criterion is defined. The algorithm is presented in Algorithm 6.4 and shown in Fig. 6.18.

6.6.2.6 Reactive Tabu Search

In the simple tabu search, the tabu list and permissible moves sets are a function of the last configuration found (local optimum). This is not a good method for finding the solution faster because the tabu list must have more elements than the permissible set when t is large. That is, permissible moves are not enough to be in the current vicinity, and the local optimum might not be the best one in that place.

An alternative to modifying the tabu tenure T is to use the reactive tabu search, which is a modification that gives the possibility of intensifying the search in regions, depending on the historical values of energy in that place. Thus, this method reacts with respect to the intensification/diversification procedures.

In this way, the simple tabu search will be the basis of the process. Then, another subroutine is defined in order to satisfy two main principles: (1) select the intensification/diversification procedure, and (2) modify the tabu tenure.

As in Table 6.1, Φ is a function that returns the number of iterations that some configuration has been visited during the searching procedure, the value is $\Phi(f)$. When the configuration $f^{(t)}$ is used, the function Π stores the actual iteration at which this configuration is evaluated, so the value recorded is $\Pi(f^{(t)}) = t$. The last function is used to store the last iteration at which the current configuration was evaluated. Suppose, that the configuration $f = f_0$ and this configuration are evaluated at time $t = t_0$. Then, five iterations later the same configuration $f^{(t+5)} = f_0$ is evaluated. The last iteration at which this configuration appeared in the searching process is then $\Pi(f^{(t+5)} = f_0) = t_0$.

Finally, this configuration comes from some move μ_i. As with the configuration, the moves have an associated function L that returns the last iteration at which the move was applied in a local optimum and the value recorded is $L(\mu_i)^{(t)} = t$. Let us suppose the same environment as the previous example. Suppose now

Algorithm 6.4	Simple tabu search
Step 1	Let the iteration time be $t = 0$. Initialize the current configuration $f^{(0)}$ randomly. The best configuration is assigned as $f_b = f^{(0)}$ and the best energy as $E_b = E(f^{(0)})$. Actually, the set of permissible moves A is all the search space and the tabu set is $\tau = \emptyset$.
Step 2	Select a criterion to decide between *intensification* (go to *Step 3*) or *diversification* (go to *Step 6*) procedures.
Step 3	With the current configuration $f^{(t)}$ do a modification in one element, the so-called $f_i^{(t)}$ running i for all permissible elements of A. If it is permitted, evaluate the energy $E(f_i^{(t)})$.
Step 4	Select the local optimum configuration temporarily known as f_l and its proper energy E_l. Then, the move μ that produces f_l must be stored in the tabu list τ.
Step 5	Compare the temporary energy with respect to the best energy. If the temporary energy is less than the best energy, then $E_b = E_l$ and $f_b = f_l$. Go to *Step 10*.
Step 6	Select another region in the permissible moves and store the elements in a subset $S \subset A$. Clear memory data used to select the intensification/diversification process.
Step 7	With the current configuration $f^{(t)}$ do a modification in one element, so-called $f_i^{(t)}$ running i for all permissible elements of S. If it is permitted, evaluate the energy $E(f_i^{(t)})$.
Step 8	Select the local optimum configuration temporarily known as f_l and its proper energy E_l. Then, the move μ that produces f_l must be stored in the tabu list τ by a period of T iterations.
Step 9	Compare the temporary energy with respect to the best energy. If the temporary energy is less than the best energy, then $E_b = E_l$ and $f_b = f_l$. Go to *Step 10*.
Step 10	Increment the iteration $t = t + 1$. Set $f^{(t+1)} = f_l$ and go to *Step 2* until the stop criterion is fired, then *STOP*.

that the configuration $f_0^{(t_0)}$ comes from the modification $\mu_x(t_0)$, then we record $L(\mu_x)^{(t_0)} = t_0$. Seven iterations later, some other configuration $f_1^{(t_0+7)}$ comes from the move μ_x, too. Then, if we need to know the last iteration at which this move was in some configuration in the searching procedure, we need to apply $L(\mu_x)^{(t_0+7)} = t_0$.

Now, the subroutine does the next few steps. At first, we evaluate the last iteration $\Pi(f^{(t)})$ at which the current configuration was evaluated and this value is assigned to a variable R. Of course, the number of iterations that this configuration has been in the searching process is updated by the rule $\Phi(f) = \Phi(f) + 1$. If the configuration is greater than some value, i.e., REP_MAX, then we store this configuration in a set of configurations named chaos C by the rule $C = C \cup f$. With this method we are able to know if we have to make a diversification procedure because the number of elements in the set C must be less than some value of threshold $CHAOS$. Otherwise, the algorithm can be in the intensification stage.

On the other hand, we need to be sure that the number of tabu elements is less than the number of permissible values. This action can be derived with the condition $R < 2(L - 1)$, which means that the number of the last iteration at which the configuration was in the searching procedure is at least double the number of the last iteration at which the move was in the process. Then, we assume that the variable R can be averaged with the equation $R_{\text{ave}} = 0.1R + 0.9R_{\text{ave}}$. This value controls the tabu tenure as shown in the Algorithm 6.5. Figure 6.19 shows the flow chart of the reactive tabu search.

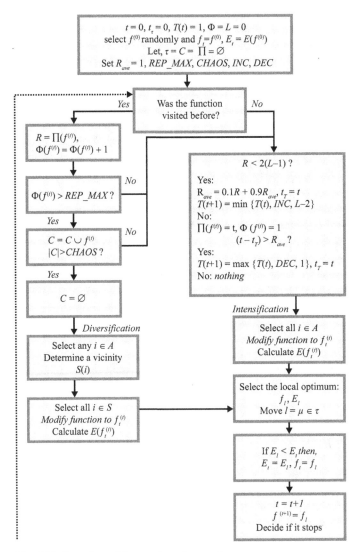

Fig. 6.19 Flow chart of the reactive tabu search

Algorithm 6.5	Reactive tabu search
Step 1	Let the iteration time be $t = 0$. Initialize the current configuration $f^{(0)}$ randomly. The best configuration is assigned as $f_b = f^{(0)}$ and the best energy as $E_b = E(f^{(0)})$. Actually, the set of permissible moves A is all the search space, and the tabu set is $\tau = \emptyset$. Let the set of chaotic configurations be $C = \emptyset$. Set the tabu tenure as $T(t) = 1$ and the last iteration at that point changed to $t_T = 0$. Set $\Pi = \emptyset$ and $\Phi = L = 0$. Initialize the variable $R_{ave} = 1$, $REP\_MAX, CHAOS, INC, DEC$.
Step 2	Evaluate $\Pi(f^{(t)})$. If there is any value set the rules: $R = \Pi(f^{(t)})$ and $\Phi(f^{(t)}) = \Phi(f^{(t)}) + 1$. Else, go to *Step 5*.
Step 3	If $\Phi(f^{(t)}) > REP\_MAX$ then update the chaotic set $C = C \cup f^{(t)}$. Else, go to *Step 5*.
Step 4	If $|C| > CHAOS$, then reinitialize $C = \emptyset$ and make a *diversification* procedure as done in *Step 10*. Else, go to *Step 5*.
Step 5	If $R < 2(L - 1)$ then do the following: $R_{ave} = 0.1R + 0.9R_{ave}$ $T(t + 1) = \min\{T(t) \cdot INC, L - 2\}$ $t_T = t$. Else, follow the next instructions: $\Pi(f^{(t)}) = t$ $\Phi(f^{(t)}) = 1$ Go to *Step 6*.
Step 6	Evaluate $(t - t_T) > R_{ave}$. If this is true then update the value of the tabu tenure $T(t + 1) = \max\{T(t) \cdot DEC, 1\}$ and record the iteration of this modification $t_T = t$ and make an *intensification* procedure as done in *Step 7*.
Step 7	With the current configuration $f^{(t)}$ do a modification in one element, so-called $f_i^{(t)}$ running i for all permissible elements of A. If it is permitted, evaluate the energy $E(f_i^{(t)})$.
Step 8	Select the local optimum configuration temporarily known as f_l and its proper energy E_l. Then, the move μ that produces f_l must be stored in the tabu list τ.
Step 9	Compare the temporary energy with respect to the best energy. If the temporary energy is less than the best energy, then $E_b = E_l$ and $f_b = f_l$. Go to *Step 14*.
Step 10	Select another region in the permissible moves and store the elements in a subset $S \subset A$. Clear memory data used to select the intensification/diversification process.
Step 11	With the current configuration $f^{(t)}$ do a modification in one element, the so-called $f_i^{(t)}$ running i for all permissible elements of S. If it is permitted, evaluate the energy $E(f_i^{(t)})$.
Step 12	Select the local optimum configuration temporarily known as f_l and its proper energy E_l. Then, the move μ that produces f_l must be stored in the tabu list τ by a period of T iterations.
Step 13	Compare the temporary energy with respect to the best energy. If temporary energy is less than the best energy, then $E_b = E_l$ and $f_b = f_l$. Go to *Step 14*.
Step 14	Increment the iteration $t = t + 1$. Set $f^{(t+1)} = f_l$ and go to *Step 2* until the stop criterion is fired, then *STOP*.

Table 6.3 Data for Example 6.3, tabu search for fuzzy associated matrices

Left	Center	Right	Motor 1	Motor 2
1	1	1	20	1
1	1	151	1	20
1	1	226	1	17
1	226	151	1	12
1	226	226	1	15
76	1	1	20	1
76	1	76	20	1
76	1	151	1	15
76	226	151	1	15
76	226	226	1	12
151	1	1	20	1
151	1	76	20	1
151	1	151	12	12
151	226	151	1	1
151	226	226	1	4
226	1	1	20	1
226	151	226	3	1
226	226	1	10	1
226	226	151	1	1
226	226	226	1	1

Example 6.5. Tabu search can be implemented in order to optimize the fuzzy associated matrix or the membership functions in fuzzy controllers. Take for example an application on robotics in which we have to optimize four input membership functions. These functions may represent the distance between the robot and some object measured by an ultrasonic sensor. We have three ultrasonic sensors measuring three distinct regions in front of the robot.

These sensors are labeled as *left*, *center* and *right*. Assume that the membership functions have the same shape for all sensors. In addition, we have experimental results in which we find values of each measure at the first three columns and the last two columns are the desired values for moving the wheels. Data is shown in Table 6.3. Use the reactive tabu search to find a good solution for the membership functions. A prototyping of those functions are shown in Fig. 6.20.

Solution. This example is implemented in LabVIEW following the path ICTL ≫ Optimizers ≫ RTS ≫ **Example RTS.vi**. The desired inputs and outputs of the fuzzy controller are already programmed inside this VI. Then, it is not necessary to copy Table 6.2.   □

We first explain the front panel. On the top-left, are the control variables. *Max Iters* is the number of times that the algorithm will be reproduced. If this number is exceeded, then the algorithm stops. The next one is *Execution Delay* that refers to a timer delay between iterations. It is just here if we want to visualize the process slowly. The *Cts* cluster has the *INC* value that controls the increment of the tabu tenure when the algorithm is evaluating if it needs an intensification or diversification process.

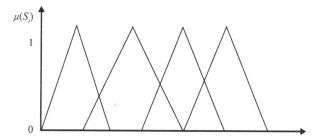

Fig. 6.20 Prototyping of membership functions

DEC is the decrement value of the tabu tenure in the same way as the last one. *CHS* is the chaos value. When the cardinality of the chaos set is greater than this value, then the algorithm makes a diversification process. Finally, the number of repetitions *REP* is known as *REP_MAX* in the algorithm previously described.

The graphs on the left side of the window are *Current f* and *Best f*. The first one shows the actual position of the membership functions. The other one shows the best configuration of the membership functions found thus far. In the middle of the window is all the information used to analyze the procedure. This cluster is called the *Information Cluster* and it is divided into three clusters.

The first is *Arr In* that shows the function f in Boolean terms (zeros and ones). *Pi, Phi* and *A* are the sets of the time at which some function was evaluated, the number of times that the function was evaluated, and the permissible moves, respectively. The *Cts* is the modified values of the control values.

Finally, on the right side of the window the *Error* graph history is shown. In this case, we have a function that determines the square of the error measured by the difference of the desired outputs and the actual outputs that the fuzzy controller returns with the actual configuration of the membership functions.

We will explain the basic steps of the reactive tabu search. First, we initialize a characterization of the configuration with 32 bits selected randomly. Of course, these bits can only have values of 0 or 1 (Fig. 6.21a). Then, we evaluate this configuration and obtain the best error thus far (Fig. 6.21b). Actually, all other values and sets explained at *Step 1* are initialized, as seen in Fig. 6.22.

Steps 2–6 in Algorithm 6.5 are known as the reaction procedure. These steps are implemented in LabVIEW in the path ICTL ≫ Optimizers ≫ RTS ≫ **rts_mbr.vi**. This VI receives three clusters: *Var In* is the cluster of the initialization values in Fig. 6.22 on the left side, *Arr In* is the cluster of the initialization values in Fig. 6.22 on the right side and *Cts In* is the cluster with the control values (*INC, DEC, CHS, REP*). Finally, it needs the configuration f. In addition, this VI returns all modified values in clusters as *Arr Out* and *Var Out*, and it determines if it needs a diversifying search procedure by the pin *Diversify?* This can be seen in Fig. 6.23.

If a diversification procedure is selected, it is implemented in the VI located at ICTL ≫ Optimizers ≫ RTS ≫ **rts_dvsm.vi**. The input connections are two clusters (*Arr In* and *Var In* explained before) and the actual configuration is known

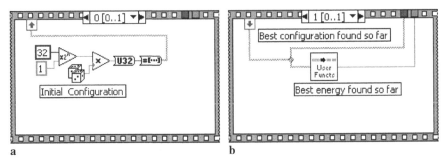

a   b

Fig. 6.21a,b Reactive tabu search implementation. **a** Initialization of the configuration with 32 bits. **b** Calculation of best error

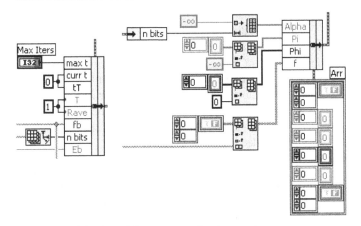

Fig. 6.22 Initialization of the parameters in tabu search

as *current f*. The output connectors are the two clusters updated (*Var Out* and *Arr Out*) and the new configuration by the pin *new f*. Figure 6.24 shows this VI.

Fig. 6.23 Determining connections of the reaction procedure

Fig. 6.24 Determining connections of the diversifying procedure

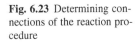

Fig. 6.25 Block diagram of the intensification procedure

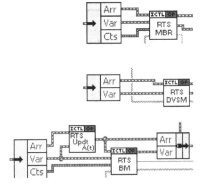

If there is no diversifying procedure, then the intensification procedure is running. This can be implemented with two VIs following the path ICTL ≫ Optimizers ≫ RTS ≫ **rts_updt-A.vi**. This VI updates the permissible moves with the information of *Arr In* and *Var In*. Then, *Arr Out* is the update of the values inside this cluster, but in fact the *A* set updating is the main purpose of this VI. The function then looks for a configuration with this permissible moves and then evaluates the best move with the VI at the path ICTL ≫ Optimizers ≫ RTS ≫ **rts_bm.vi**. This VI takes *Arr In*, *Var In* and the actual configuration *current f*. This returns *Var Out*

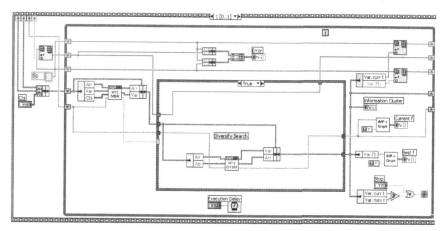

Fig. 6.26 Block diagram of the RTS example

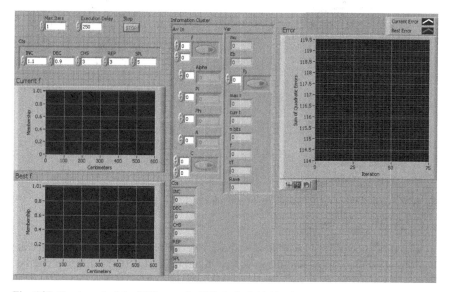

Fig. 6.27 Front panel of the RTS example. This is the initialization step

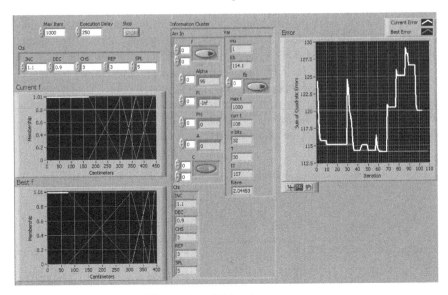

Fig. 6.28 Front panel of the RTS example at 100 iterations

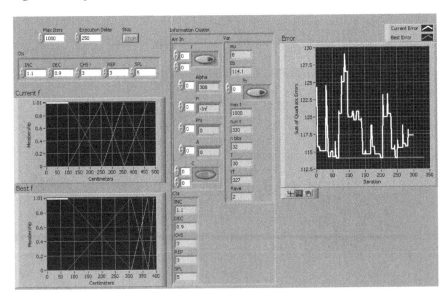

Fig. 6.29 Front panel of the RTS example at 300 iterations

(as a modification procedure of these values), *Energy* that is the energy evaluated at the current configuration, and the *new f* configuration. This block diagram can be viewed in Fig. 6.25.

After either intensification or diversification procedures, we have to choose if the configuration is better than the best configuration found thus far. It is easy to com-

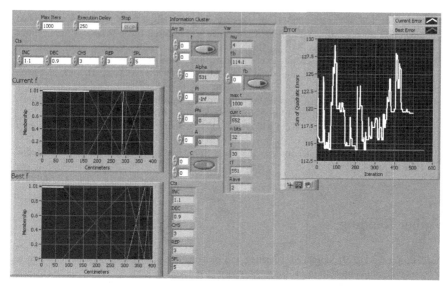

Fig. 6.30 Front panel of the RTS example at 500 iterations

pare the best configuration up to this point that comes from the *Var In* cluster and the actual configuration coming from any of the searching procedures. Figure 6.26 shows a global visualization of the block diagram of this example.

Continuing with the example, let $INC = 1.1$, $DEC = 0.9$, $CHS = 3$ and $REP = 3$. The maximum number of iterations $Max\ Iters = 1000$ and $Execution\ Delay = 250$. Finally, we can look at the behavior of this optimization procedure in Fig. 6.26 (initialization) and Figs. 6.27–6.30. As we can see, at around 80 iterations, the best solution was found.

References

1. Khouja M, Booth DE (1995) Fuzzy clustering procedure for evaluation and selection of industrial robots. J Manuf Syst 14(4):244–251
2. Saika S, et al. (2000) WSSA: a high performance simulated annealing and its application to transistor placement. Special Section on VLSI Design and CAD Algorithms. IEICE Trans Fundam Electron Commun Comput Sci E83-A(12):2584–2591
3. Bailey RN, Garner KM, Hobbs MF (1997) Using simulated annealing and genetic algorithms to solve staff-scheduling problems. Asia Pacific J Oper Res Nov 1 (1997)
4. Baker JA, Henderson D (1976) What is liquid? Understanding the states of matter. Rev Mod Phys 48:587–671
5. Moshiri B, Chaychi S (1998) Identification of a nonlinear industrial process via fuzzy clustering. Lect Notes Comput Sci 1415:768–775
6. Zhang L, Guo S, Zhu Y, Lim A (2005) A tabu search algorithm for the safe transportation of hazardous materials. Symposium on Applied Computing, Proceedings of the 2005 ACM Symposium on Applied Computing, Santa Fe, NM, pp 940–946

7. Shuang H, Liu Y, Yang Y (2007) Taboo search algorithm based ANN model for wind speed prediction. Proceedings of 2nd IEEE Conference on Industrial Electronics and Applications (ICIEA 2007), 23–25 May 2007, pp 2599–2602
8. Brigitte J, Sebbah S (2007) Multi-level tabu search for 3G network dimensioning. Proceedings of IEEE Wireless Communications and Networking Conference (WCNC 2007), 11–15 March 2007, pp 4411–4416
9. Lee CY, Kang HG (2000) Cell planning with capacity expansion in mobile communications: a tabu search approach. IEEE Trans Vehic Technol 49(5):1678–1691
10. Kirkpatrick S, Gelatt CD, Vecchi MP (1983) Optimization by simulated annealing. Science 220:671–680
11. Jones MT (2003) AI application programming. Simulated Annealing. Charles River Media, Boston, pp 14–33

Futher Reading

Aarts E, Korst J (1989) Simulated annealing and Boltzmann machines: a stochastic approach to combinatorial optimization and neural computing. Wiley, New York

Battiti R, Tecchiolli G (1994) The reactive tabu search. ORSA J Comput 6(2):126–140

Dumitrescu D, Lazzerini B, Jain L (2000) Fuzzy sets and their application to clustering and training. CRC, Boca Raton, FL

Glover F, Laguna M (2002) Tabu search. Kluwer Academic, Dordrecht

Klir G, Yuan B (1995) Fuzzy sets and fuzzy logic. Prentice Hall, New York

Reynolds AP, et al. (1997) The application of simulated annealing to an industrial scheduling problem. Proceedings on Genetic Algorithms in Engineering Systems: Innovations and Applications, 2–4 Sept 1997, No 446, pp 345–350

Windham MP (1981) Cluster validity for fuzzy clustering algorithms. Fuzzy Sets Syst 5:177–185

Chapter 7
Predictors

7.1 Introduction to Forecasting

Predictions of future events and conditions are called *forecasts*; the act of making predictions is called *forecasting*. Forecasting is very important in many organizations since predictions of future events may need to be incorporated in the decision-making process. They are also necessary in order to make intelligent decisions. A university must be able to forecast student enrollment in order to make decisions concerning faculty resources and housing availability.

In forecasting events that will occur in the future, a forecaster must rely on information concerning events that have occurred in the past. That is why the forecasters must analyze past data and must rely on this information to make a decision. The past data is analyzed in order to identify a pattern that can be used to describe it. Then the pattern is extrapolated or extended to forecast future events. This basic strategy is employed in most forecasting techniques rest on the assumption that a pattern that has been identified will continue in the future.

Time series are used to prepare forecasts. They are chronological sequences of observations of a particular variable. Time series are often examined in hopes of discovering a historical pattern that can be exploited in the preparation of a forecast. An example is shown in Table 7.1.

Table 7.1 Data for forecasting example

Time [s]	Current [mA]
0.1	1.1
0.2	0.9
0.3	0.8
0.4	0.65
0.5	0.45

A time series is a composition of several components, in order to identify patterns:

1. *Trend*. Refers to the upward or downward movement that characterizes a time series over a period of time. In other words, it reflects the long-run growth or decline in the time series.
2. *Cycle*. Recurring up and down movements around trend levels.
3. *Seasonal variations*. Periodic patterns in time series that complete themselves within a period and are then repeated on that basis.
4. *Irregular fluctuations*. Erratic movements in a time series that follow no recognizable or regular patterns. These movements represent what is left over in a time series after the other components have been accounted for. Many of these fluctuations are caused by unusual events that cannot be forecasted.

These components do not always occur alone, they can occur in any combination or all together, for this reason no single best forecasting model exists. Thus, one of the most important problems to be solved in forecasting is that of trying to match the appropriate model to the model of the available time series data.

7.2 Industrial Applications

Predictors or forecasters are very useful in the industry. Some applications related to this topic are summarized in the following:

Stock index prediction. Companies or governments need to know about their resources in stock. This is why predictors are constantly used in those places. In general, they are looking for some patterns about the potential market and then they have to offer their products. In these terms, they want to know how many products could be offered in the next few months. Statistically, this is possible with predictors or forecasters knowing the behavior of past periods. For example, Shen [1] reports a novel predictor based on gray models using some neural networks. Actually, this model was used to predict the monetary changes in Shanghai in the years 2006 and 2007. Other applications in stock index forecasting are reported in [1].

Box–Jenkins forecasting in Singapore. Dealing with construction industry demand, Singapore needed to evaluate the productivity of this industry, its construction demand, and tend prices in the year 2000. This forecasting was applied with a Box–Jenkins model. The full account of this approach researched by the School of Building and Real Estate, National University of Singapore is found in the work by B.H. Goa and H.P. Teo [2].

Pole assignment controller for practical applications. In the industry, controllers are useful in automated systems, industry production, robotics, and so on. In these terms, a typical method known as generalized minimum variance control (GMVC) is used that aims to self-tune its parameters depending on the application. However, this method is not implemented easily. In Mexico, researchers designed a practical

GMVC method in order to make it feasible [3]. They used the minimum variance control technique to achieve this.

Inventory control. In the case of inventory control, exponential smoothing forecasters are commonly used. As an example of this approach, Snyder et al. published a paper [4] in which they describe an inventory management of seasonal product of jewelry.

Dry kiln transfer function. In a control field, the transfer function is an important part of the designing and analyzing procedures. Practical applications have non-linear relations between their input and output variables. However, transfer functions cannot be applied in that case because it has an inherent linear property. Forecasting is then used to set a function of linear combinations in statistical parameters. Blankenhorn et al. [5] implemented a Box–Jenkins method in the transfer function estimations. Then, classical control techniques could be applied. In Blankenhorn's application, they controlled a dry kiln for a wood drying process.

7.3 Forecasting Methods

The two main groups in which forecasting techniques can be divided are qualitative methods and quantitative methods; they will be further described in the following section.

7.3.1 Qualitative Methods

They are usually subject to the opinion of experts to predict future events. These methods are usually necessary when historical data is not available or is scarce. They are also used to predict changes in historical data patterns. Since the use of historical data to predict future events is based on the assumption that the pattern of the historical data will persist, changes in the data pattern cannot be predicted on the basis of historical data. Thus, qualitative methods are used to predict such changes.

Some of these techniques are:

1. *Subjective curve fitting.* Depending on the knowledge of an expert a curve is built to forecast the response of a variable, thus this expert must have a great deal of expertise and judgment.
2. *Delphi method.* A group of experts is used to produce predictions concerning a specific question. The members are physically separated, they have to respond to a series of questionnaires, and then subsequent questionnaires are accompanied by information concerning the opinions of the group. It is hoped that after several rounds of questions the group's response will converge on a consensus that can be used as a forecast.

7.3.2 Quantitative Methods

These techniques involve the analysis of historical data in an attempt to predict future values of a variable of interest. They can be grouped into two kinds: *univariate* and *causal* models.

The univariate model predicts future values of a time series by only taking into account the past values of the time series. Historical data is analyzed attempting to identify a data pattern, and then it is assumed that the data will continue in the future and this pattern is extrapolated in order to produce forecasts. Therefore they are used when conditions are expected to remain the same.

Casual forecasting models involve the identification of other variables related to the one to be predicted. Once the related variables have been identified a statistical model describing the relationship between these variables and the variable to be forecasted is developed. The statistical model is used to forecast the desired variable.

7.4 Regression Analysis

Regression analysis is a statistical methodology that is used to relate variables. The variable of interest or dependent variable (y) that we want to analyze is to be related to one or more independent or predictive variables (x). The objective then is to use a regression model and use it to describe, predict or control the dependent variables on the basis of the independent variables.

Regression models can employ *quantitative* or *qualitative* independent variables. Quantitative independent variables assume numerical values corresponding to points on the real line. Qualitative independent variables are non-numerical. The models are then developed using observed models of the dependent and independent variables. If these values are observed over time, the data is called a *time series*. If the values are observed at one point in time, the data are called *cross-sectional data*.

7.5 Exponential Smoothing

Exponential smoothing is a forecasting method that weights the observed time series values *unequally* because more recent observations are weighted more heavily than more remote observations. This unequal weighting is accomplished by one or more *smoothing constants*, which determine how much weight is given to each observation. It has been found to be most effective when the parameters describing the time series may be changing *slowly over time*.

Exponential smoothing methods are not based on any formal model or theory; they are techniques that produce adequate forecasts in some applications. Since these techniques have been developed without a theoretical background some practitioners strongly object to the term model in the context of exponential smoothing. This method assumes that the time series has no trend while the level of the time series may change slowly over time.

7.5.1 Simple-exponential Smoothing

Suppose that a time series is appropriately described by the no trend equation: $y_t = \beta_0 + \varepsilon_t$. When β_0 remains constant over time it is reasonable to forecast future values of y_t by using regression analysis. In such cases the least squares point estimate of β_0 is

$$b_0 = \bar{y} = \sum_{t=1}^{n} \frac{y_t}{n} .$$

When computing the point estimate b_0 we are equally weighting each of the previous observed time series values of $y_1, \ldots, y_n$. When the value of β_0 slowly changes over time, the equal weighting scheme may not be appropriate. Instead, it may be desirable to weight recent observations more heavily than remote observations. Simple-exponential smoothing is a forecasting method that applies unequal weights to the time series observations. This is accomplished by using a smoothing constant that determines how much weight is given to the observation.

Usually the most recent is given the most weight, and older observations are given successively smaller weights. The procedure allows the forecaster to update the estimate of β_0 so that changes in the value of this parameter can be detected and incorporated into the forecasting system.

7.5.2 Simple-exponential Smoothing Algorithm

1. The time series $y_1, \ldots, y_n$ is described by the model $y_t = \beta_0 + \varepsilon_t$, where the average level β_0 may be slowly changing over time. Then the estimate $a_0(T)$ of β_0 made in time period T is given by the smoothing equation:

$$a_0 = \alpha y_T + (1 - \alpha) a_0 (T - 1) , \qquad (7.1)$$

where α is the smoothing constant between 0 and 1 and $a_0(T-1)$ is the estimate of β_0 made in time period $T-1$.
2. A point forecast or one-step-ahead forecast made in time period T for $y_{T+\tau}$ is:

$$\hat{y}_{T+\tau}(T) = a_0(T) . \qquad (7.2)$$

3. A $100(1 - \alpha)\%$ prediction interval computed in time period T for $y_{T+\tau}$ is:

$$\left[a_0(T) \pm z_{[\alpha/2]} 1.25 \Delta(T)\right] , \qquad (7.3)$$

where $\Delta(T) = \dfrac{\sum\limits_{t=1}^{T} [y_t - a_0(T-1)]}{T}$.
4. If we observe y_{T+1} in the time period $T + 1$, we can update $a_0(T)$ and $\Delta(T)$ to $a_0(T+1)$ and $\Delta(T+1)$ by:

$$a_0(T+1) = \alpha y_{T+1} + (1 - \alpha) a_0(T) \qquad (7.4)$$

$$\Delta(T+1) = \frac{T\Delta(T) + [y_{T+1} - a_0(T)]}{T + 1} . \qquad (7.5)$$

Therefore a point forecast made in time period $T + 1$ for $y_{T+1+\tau}$ is:

$$\left[a_0 \left(T + 1\right) \pm z_{[\alpha/2]} 1.25 \Delta \left(T + 1\right)\right] . \tag{7.6}$$

7.5.2.1 Adaptation of Parameters

Sometimes it is necessary to change the smoothing constants being employed in exponential smoothing. The decision to change smoothing constants can be made by employing *adaptive control procedures.* By using a tracking signal we will have better results in the forecasting, by realizing that the forecast error is larger than an accurate forecasting system might reasonably produce.

We will suppose that we have accumulated the T single-period-ahead forecast errors $e_1\left(\alpha\right), \ldots, e_T\left(\alpha\right)$, where α denotes the smoothing value used to obtain a single-step-ahead forecast error. Next we define the sum of these forecast errors: $Y\left(\alpha, T\right) = \sum_{t=1}^{T} e_t\left(\alpha\right)$. With that we will have $Y\left(\alpha, T\right) = Y\left(\alpha, T-1\right) + e_T\left(\alpha\right)$; and we define the following mean absolute deviation as:

$$D\left(\alpha, T\right) = \frac{\sum\limits_{t=1}^{T} |e_t\left(\alpha\right)|}{T} . \tag{7.7}$$

Then the *tracking* signal is defined as:

$$TS\left(\alpha, T\right) = \left|\frac{Y\left(\alpha, T\right)}{D\left(\alpha, T\right)}\right| . \tag{7.8}$$

So when $TS\left(\alpha, T\right)$ is large it means that $Y\left(\alpha, T\right)$ is large relative to the mean absolute deviation of $D\left(\alpha, T\right)$. By that we understand that the forecasting system is producing errors that are either consistently positive or negative. It is a good measure of an accurate forecasting system to produce one-half positive errors and one-half negative errors.

Several possibilities exist if the tracking system indicates that correction is needed. Variables may be added or deleted to obtain a better representation of the time series. Another possibility is that the model used does not need to be altered, but the parameters of the model need to be. In the case of exponential smoothing, the constants would have to be changed.

7.5.3 Double-exponential Smoothing

A time series could be described by the following linear trend: $y_t = \beta_0 + \beta_1 t + \varepsilon_t$.

When the values of the parameters β_0 and β_1 slowly change over the time, *double-exponential smoothing* can be used to apply unequal weightings to the time series observations. There are two variants of this technique: the first one employs one smoothing constant. It is often called *one-parameter double-exponential smoothing*. The second is the *Holt–Winters two-parameter double-exponential smoothing*, which employs two smoothing constants. The smoothing constants determine how much weight is given to each time series observation.

The one-parameter double-exponential smoothing employs *single and double-smoothed statistics*, denoted as S_T and $S_T^{[2]}$. These statistics are computed by using two smoothing equations:

$$S_T = \alpha y_t + (1 - \alpha) S_{T-1} \tag{7.9}$$

$$S_T^{[2]} = \alpha S_t + (1 - \alpha) S_{T-1}^{[2]} . \tag{7.10}$$

Both of these equations use the same smoothing constant α, defined between 0 and 1. The first equation smoothes the original time series observations; the second smoothes the S_T values that are obtained by using the first equation. The following estimates are obtained as shown:

$$b_1 (T) = \frac{\alpha}{1 - \alpha} \left(S_T - S_T^{[2]} \right) \tag{7.11}$$

$$b_0 (T) = 2 S_T - S_T^{[2]} - T b_1 (T) . \tag{7.12}$$

With the estimates $b_1 (T)$ and $b_0 (T)$, a forecast made at time T for the future value $y_{T+\tau}$ is:

$$\hat{y}_{T+\tau} (T) = b_0 (T) + b_1 (T) (T + \tau) = [b_0 (T) + b_1 (T) T] + b_1 (T) \tau$$
$$= a_0 (T) + b_1 (T) \tau . \tag{7.13}$$

where $a_0 (T)$ is an estimate of the updated trend line with the time origin considered to be at time T. That is, $a_0 (T)$ is the estimated intercept with time origin considered to be at time 0 plus the estimated slope multiplied by T. It follows:

$$a_0 (T) = b_0 (T) + b_1 (T) T = \left[2 S_T - S_T^{[2]} - T b_1 (T) \right] + b_1 (T) T = 2 S_T - S_T^{[2]} . \tag{7.14}$$

Finally the forecast of $y_{T+\tau} (T)$ is:

$$\hat{y}_{T+\tau} (T) = a_0 (T) + b_1 (T) \tau = 2 S_T - S_T^{[2]} + \frac{\alpha}{1 - \alpha} \left(S_T - S_T^{[2]} \right) \tau$$
$$= \left(2 + \frac{\alpha \tau}{1 - \alpha} \right) S_T - \left(1 + \frac{\alpha \tau}{1 - \alpha} \right) S_T^{[2]} . \tag{7.15}$$

7.5.4 Holt–Winter Method

This method is widely used on adaptive prediction and predictive control applications. It is simple yet a robust method. It employs two smoothing constants. Suppose that in time period $T - 1$ we have an estimate $a_0 (T - 1)$ of the average level of the time series. In other words, $a_0 (T - 1)$ is an estimate of the intercept of the time series when the time origin is considered to be time period $T - 1$.

If we observe y_T in time period T, then:

1. The updated estimate $a_0 (T)$ of the permanent component is obtained by:

$$a_0 (T) = \alpha y_T + (1 - \alpha) [a_0 (T - 1) + b_1 (T - 1)] . \qquad (7.16)$$

Here α is the smoothing constant, which is in the range $[0, 1]$.
2. An updated estimate is $b_1 (T)$ if the trend component is obtained by using the following equation:

$$b_1 (T) = \beta [a_0 (T) - a_0 (T - 1)] + (1 - \beta) b_1 (T - 1) , \qquad (7.17)$$

where β is also a smoothing constant, which is in the range $[0, 1]$.
3. A point forecast of future values of $y_{T+\tau} (T)$ at time T is: $y_{T+\tau} (T) = a_0 (T)$ $+ b_1 (T) \tau$.
4. Then we can calculate an approximate $100 (1 - \alpha)\%$ prediction interval for $y_{T+\tau} (T)$ as $[\hat{y}_{T+\tau} (T) \pm z_{\alpha/2} d_\tau \Delta(T)]$, where d_τ is given by:

$$d_\tau = 1.25 \left[\frac{1 + \frac{\theta}{(1+v)^3} \left[(1 + 4v + 5v^2) + 2\theta (1 + 3v) \tau + 2\theta^2 \tau^2 \right]}{1 + \frac{\theta}{(1+v)^3} \left[(1 + 4v + 5v^2) + 2\theta (1 + 3v) \tau + 2\theta^2 \right]} \right] .$$
$$(7.18)$$

Here θ equals the maximum of α and β, $v = 1 - \theta$, and

$$\Delta (T) = \frac{\sum_{t=1}^{T} |y_t - [a_0 (t - 1) + b_1 (t - 1)]|}{T} . \qquad (7.19)$$

5. Observing y_{T+1} in the time period $T + 1$, $\Delta (T)$ may be updated to $\Delta (T + 1)$ by the following equation:

$$\Delta (T + 1) = \frac{T\Delta (T) + |y_{T+1} - [a_0 (T) + b_1 (T)]|}{T + 1} . \qquad (7.20)$$

7.5.5 Non-seasonal Box–Jenkins Models

The classical Box–Jenkins model describes a *stationary* time series. If the series that we want to forecast is not stationary we must transform it into one. We say that a time series is *stationary* if the statistical properties like *mean* and *variance* are constant through time. Sometimes the non-stationary time series can be transformed into stationary time series values by taking the first differences of the non-stationary time series values.

This is done by: $z_t = y_t - y_{t-1}$ where $t = 2, \ldots, n$. From the experience of experts in the field, if the original time series values $y_1, \ldots, y_n$ are *non-stationary* and *non-seasonal* then using the first differencing transformation $z_t = y_t - y_{t-1}$ or the second differencing transformation $z_t = (y_t - y_{t-1}) - (y_{t-1} - y_{t-2}) = y_t - 2y_{t-1} + y_{t-2}$ will usually produce stationary time series values.

Once the original time series has been transformed into stationary values the Box–Jenkins model must be identified. Two useful models are *autoregressive* and *moving average* models.

Moving average model. The name refers to the fact that this model uses past random shocks in addition to using the current one: $a_t, a_{t-1}, \ldots, a_{t-q}$. The model is given as:

$$z_t = \delta + a_t - \theta_1 a_{t-1} - \theta_2 a_{t-2} - \cdots - \theta_q a_{t-q} . \tag{7.21}$$

Here the terms $\theta_1, \ldots, \theta_n$ are unknown parameters relating z_t to $a_{t-1}, a_{t-2}, \ldots, a_{t-q}$. Each random shock a_t is a value that is assumed to be randomly selected from a normal distribution, with a mean of zero and the same variance for each and every time period. They are also assumed to be statistically independent.

Autoregressive model. The model $z_t = \delta + \phi_1 z_{t-1} + \ldots + \phi_p z_{t-p} + a_t$ is called the *non-seasonal autoregressive model of order p*. The term autoregressive refers to the fact that the model expresses the current time series value z_t as a function of past time series values $z_{t-1} + \ldots + z_{t-p}$. It can be proved that for the non-seasonal autoregressive model of order p that:

$$\delta = \mu \left(1 - \phi_1 - \phi_2 - \cdots - \phi_p\right) . \tag{7.22}$$

7.5.6 General Box–Jenkins Model

In the previous section, Box–Jenkins offers a description of a non-seasonal time series. Now, it can be rephrased in order to find a forecasting of seasonal time series. This discussion will introduce the general notation of stationary transformations.

Let B be the *backshift operator* defined as $By_t = y_{t-1}$ where y_i is the ith time series observation. This means that B is an operator under the ith observation in order to get the $(i-1)$th observation. Then, the operator B^k refers to the $(i - k)$th time series observation like $B^k y_t = y_{t-k}$.

Then, a *non-seasonal operator* ∇ is defined as $\nabla = 1 - B$ and the *seasonal operator* ∇_L is $\nabla_L = 1 - B^L$, where L is the number of seasons in a year (measured in months).

In this case, if we have either a pre-differencing transformation $y_t^* = f(y_t)$, where any function f or not like $y_t^* = y_t$, then a *general stationary transformation* is given by:

$$z_t = \nabla_L^D \nabla^d y_t^* \tag{7.23}$$

$$z_t = (1 - B^L)^D (1 - B)^d y_t^* , \tag{7.24}$$

where D is the degree of seasonal differencing and d is the degree of non-seasonal differencing. In other words, it refers to the fact that the transformation is proportional to a seasonal differencing times a non-seasonal differencing.

We are ready to introduce the generalization of the Box–Jenkins model. We say that the Box–Jenkins model has order (p, P, q, Q) if it is: $\phi_p(B)\phi_P(B^L)z_t = \delta + \theta_q(B)\theta_Q(B^L)a_t$. Then, this is called the generalized Box–Jenkins model of order (p, P, q, Q), where:

- $\phi_p(B) = (1 - \phi_1 B - \phi_2 B^2 - \cdots - \phi_p B^p)$ is called the *non-seasonal autoregressive operator of order p.*
- $\phi_P(B^L) = (1 - \phi_{1,L} B^L - \phi_{2,L} B^{2L} - \cdots - \phi_{P,L} B^{PL})$ is called the *seasonal autoregressive operator of order P.*
- $\theta_q(B) = (1 - \theta_1 B - \theta_2 B^2 - \cdots - \theta_q B^q)$ is called the *non-seasonal moving average operator of order q.*
- $\theta_Q(B^L) = (1 - \theta_{1,L} B^L - \theta_{2,L} B^{2L} - \cdots - \theta_{Q,L} B^{QL})$ is called the *seasonal moving average operator of order Q.*
- $\delta = \mu\phi_p(B)\phi_P(B^L)$ in which μ is the true mean of the stationary time series being modeled.
- All terms $\phi_1, \ldots, \phi_p, \phi_{1,L}, \ldots, \phi_{P,L}, \theta_1, \ldots, \theta_q, \theta_{1,L}, \ldots, \theta_{Q,L}, \delta$ are unknown values that must be estimated from sample data.
- $a_t, a_{t-1}, \ldots$ are random shocks assumed statically independent and randomly selected from a normal distribution with mean value zero and variance equal for each and every time period t.

7.6 Minimum Variance Estimation and Control

It can be defined in statistics that a uniformly minimum variance estimator is an estimator with a lower variance than any other unbiased estimator for all possible values of the parameter. If an unbiased estimator exists, it can be proven that there is an essentially unique estimator.

A minimum variance controller is based on the minimum variance estimator. The aim of the standard minimum variance controller is to regulate the output of a stochastic system to a constant set point. We can express it in optimization terms in the following.

For each period of time t, choose the control $u(t)$ that will minimize the output variance:

$$J = E\left[y^2(t + k)\right], \tag{7.25}$$

where k is the time delay. The cost junction J involves k because $u(t)$ will only affect $y(s)$ for $s \geq t + k$. J will have the same minimum value for each t (asymptotically) if the controller leads to a closed-loop stability and the output is a stationary process.

The difference equation has the form $y(t) = ay(t - 1) + au(t - 1) + e(t) + ce(t - 1)$, where $e(t)$ is zero mean white noise of variance σ_e^2. If $k = 1$ then we will have:

$$y(t+1) = ay(t) + bu(t) + e(t+1) + ce(t) . \qquad (7.26)$$

Independently from the choice of the controller, $u(t)$ cannot physically be a function of $y(t+1)$, so that $\hat{y}(t+1|t)$ is functionally independent of $e(t+1)$. Then we form the J cost function as:

$$J = E\left[y^2(t+1)\right] = E\left[\hat{y}(t+1|t) + e(t+1)\right]^2 =$$
$$E\left[\hat{y}(t+1|t)\right]^2 + E\left[e(t+1)\right]^2 + 2E\left[\hat{y}(t+1|t)e(t+1)\right] . \qquad (7.27)$$

Then we can assume that the right-hand side vanishes for: (a) any linear controller, and (b) any non-linear controller, provided $e(t)$ is an independent sequence (not just uncorrelated). We know that condition (b) is satisfied by assuming a white common noise. This will reduce the cost function to: $J = E\left[\hat{y}(t+1|t)\right] + \sigma_e^2$.

Therefore J can be minimized if $u(t)$ can be chosen to satisfy $\hat{y}(t+1|t) = ay(t) + bu(t) + ce(t) = 0$. The question arises as to what gives us an implementable control law if $e(t)$ can only be expressed as a function of available data, which can be achieved by the process equation $e(t) = y(t) - ay(t-1) - bu(t-1) - ce(t-1)$. This function can be expressed in transfer function terms as:

$$e(t) = \frac{1}{1 + cz^{-1}}\left[\left(1 - az^{-1}\right)y(t) - bz^{-1}u(t)\right] . \qquad (7.28)$$

Recursion always requires unknown initial values of the noise signal unless c is zero. This reconstruction of $e(t)$ is only valid asymptotically with $|c| < 1$. This last condition is weak for processes that are stationary and stochastic. We can write $\hat{y}(t+1|t)$ with the aid of $e(t)$ in is transfer function as:

$$\hat{y}(t+1|t) = \frac{1}{1 + cz^{-1}}\left[(a+c)y(t) + bu(t)\right] . \qquad (7.29)$$

If we set $\hat{y}(t+1|t)$ to zero it will yield to a minimum variance (MV) regulator:

$$u(t) = -\frac{a+c}{b}y(t) . \qquad (7.30)$$

Rewriting some equations as $y(t+1) = \hat{y}(t+1|t) + e(t+1)$, the closed-loop behavior under $u(t)$ is then given by $y(t+1) = e(t+1)$. With this the minimum achievable variance is σ_e^2, but it will not happen if the time delay is greater than unity.

From the previous equations we can see that the developed control law exploits the noise structure of the process. Returning to the equation $y(t+1) = \hat{y}(t+1|t) + e(t+1)$, we note that $y(t+1)$ is the sum of two independent terms. The first is a function of data up to time t with the minimum achievable output variance $\sigma_e^2 = E\left[y(t+1) - \hat{y}(t+1|t)\right]^2$. We find that $e(t+1)$ cannot be reconstructed from the available data. That is why we can interpret $\hat{y}(t+1|t)$ as the best possible estimate at time t.

A more general framework to minimize the cost function could be with a CARMA model. $Ay(t) = z^{-k}Bu(t) + Ce(t)$, so we have $y(t+k) = \frac{B}{A}u(t) + \frac{C}{A}e(t+k)$. Now we must define the polynomials F, G that will satisfy the equation for $C = AF + z^{-k}G$:

$$F = 1 + f_1 z^{-1} + \ldots + f_{k-1}z^{-(k-1)}$$
$$G = g_0 + g_1 z^{-1} + \ldots + g_{n_g}z^{-n_g}$$
$$n_g = \max(n_a - 1, n_c - k), \qquad (7.31)$$

where F will represent the first k terms in the expansion of C/A. After developing the equations a little we will have:

$$y(t+k) = \left[\frac{BF}{C}u(t) + \frac{G}{C}y(t)\right] + Fe(t+k). \qquad (7.32)$$

where the first term $\hat{y}(t+k|t) = \left[\frac{BF}{C}u(t) + \frac{G}{C}y(t)\right]$ is considered the best prediction given at time t. The output prediction error is $Fe(t+k) = y(t+k) - \hat{y}(t+k|t)$, which arises from the signals $e(t+1), \ldots, e(t+k)$. These errors cannot be eliminated by $u(t)$. The cost function will be of the form:

$$J = E\left[y^2(t+k)\right] = E[\hat{y}(t+k|t) + Fe(t+k)]^2$$
$$= E[\hat{y}(t+k|t)]^2 + \left(1 + f_1^2 + \ldots + f_{k-1}^2\right)\sigma_e^2, \qquad (7.33)$$

which can be minimized by the predicted output set equal to zero. This will yield the following control law of $BFu(t) + Gy(t) = 0$ and the output signal $y(t) = Fe(t)$. This will correspond to the minimum output variance $J_{min} = \left(1 + f_1^2 + \ldots + f_{k-1}^2\right)\sigma_e^2$.

7.7 Example of Predictors Using the Intelligent Control Toolkit for LabVIEW (ICTL)

We will now create a program that will contain the exponential smoothing, Box–Jenkins model, and minimum variance predictors. We will briefly explain the equations and how they are programmed.

7.7.1 Exponential Smoothing

This is one of the most popular methods, based on time series and transfer function models. It is simple and robust, where the time series are modeled through a low pass filter. The signal components may be individually modeled, like trend, average, periodic component, among others.

The exponential smoothing is computationally simple and fast, while at the same time this method can perform well in comparison with other complex methods [6]. These methods are principally based on the heuristic understanding of the underlying process, and both time series with and without seasonality may be treated.

A popular approach for series without seasonality is the Holt method. The series used for prediction is considered a composition of more than one structural component (average and trend), each of which can be individually modeled. Such type of series can be expressed as: $y(x) = y_{av}(x) + py_{tr}(x) + e(x)$; $p = 0$ [7, 8], where $y(x)$, $y_{av}(x)$, $y_{tr}(x)$, and $e(x)$ are the data, the average, the trend and the error components individually modeled using exponential smoothing. The p-step-ahead prediction is given by $y^* (x + p|k) = y_{av}(x) + py_{tr}(x)$.

The average and the trend components are modeled as:

$$y_{av}(x) = (1 - \alpha) y(x) + \alpha (y_{av}(x - 1) + y_{tr}(k - 1)) \qquad (7.34)$$
$$y_{tr}(x) = (1 - \beta) y_{tr}(x - 1) + \beta (y_{av}(x) + y_{av}(x - 1)) , \qquad (7.35)$$

where α and β are the smoothing coefficients, whose values can be between $(0, 1)$; typical values range from 0.1 to 0.3 [8, 9]. The terms y_{av} and y_{tr} were initialized as:

$$y_{av}(1) = y(1) \qquad (7.36)$$
$$y_{tr}(1) = \frac{(y(1) - y(0)) + (y(2) - y(1))}{2} . \qquad (7.37)$$

7.7.2 Box–Jenkins Method

This is one of the most powerful methods of prediction, where the data structures are transformed and converted to stationary series represented by a transfer function model. The computational requirements are moderately high but it has been successfully applied to a variety of processes. It involves essentially two elements [10]:

1. Transformation of the time series into stationary time series.
2. Modeling and prediction of the transformed data using a transfer function model.

A discrete-time linear model of the time series is used. The series are transformed into stationary series to ensure that the probabilistic properties of mean and variance remain invariant over time.

The process is modeled as a liner filter driven by a white noise sequence. A generalized model can be expressed as $A (q^{-1}) y (k) = C (q^{-1}) e (k)$, where:

$$A (q^{-1}) = 1 + a_1 q^{-1} + \ldots + a_p q^{-p}$$
$$C (q^{-1}) = 1 + c_1 q^{-1} + \ldots + c_r q^{-s} .$$

The term $\{e(k)\}$ is a discrete white noise sequence and $\{y(k)\}$ is the time series. The backward shift operator is expressed as q^{-1}. Before the data series can be used for modeling they may be subjected to non-linear and stationary transformation.

The dth-order differencing for non-seasonal time-differencing is given by $Y_d(k) = (1 - q^{-1})^d y(k)$, which results in d successive time differences being performed on the data. A generalized model is given by $A(q^{-1}) \Delta^d y(k) = C(q^{-1}) e(k)$.

This is known as an autoregressive integrated moving average (ARIMA) model of order (p, q, r). The p are the autoregressive terms, d is the degree of time differences, and r is the order of the moving average, where the discrete time polynomials are of order p and r, respectively.

A one-step-ahead minimum mean square error prediction is the conditional expectation of $y(k + p)$ at time k: $y^{\wedge}(k + 1 | k) = E(y(k + 1) | y(k), y(k + 1) \ldots)$.

The error sequence may be expressed as $e(k) = y(k) - y^{\wedge}(k | k - 1), \ldots$ Once the parameters are estimated the predictions can be computed. The prediction of an ARIMA$(1, 1, 1)$ process, considers the model:

$$\left(1 - a_1 q^{-1}\right) \Delta y(k) = \left(1 - c_1 q^{-1}\right) e(k) . \tag{7.38}$$

The error is the difference between the real value and the prediction. A one-step-ahead prediction is given by $y^{\wedge}(k + 1 | k) = (1 + a_1) y(k) - a_1 y(k - 1) - c_1 e(k)$, where a_1 and c_1 are the estimated parameter values.

7.7.3 Minimum Variance

This kind of predictor takes the variance of the prediction error σ_e^2, as a measure of the trust in the prediction [11]. A one-step predictor can be obtained considering the process $y(t) = ay(t - 1) + e(t) + ce(t - 1)$, such as $A = 1 - az^{-1}$, $C = 1 + cz^{-1}$. For the one-step predictor $k = 1$, $F = 1 : z^{-1}G = C - A = (c + a)z^{-1}$ and

$$G(z) = c + a \text{ so: } y^*(t + 1 | t) = \left[\frac{c + a}{1 + cz^{-1}}\right] y(t) .$$

Expressed recursively gives $y^*(t + 1 | t) = (c + a) y(t) - cy^*(t | t - 1)$.

No we will program a double-exponential smoothing prediction system using the ICTL. We can find the predictor VIs at the *Predictors* palette, as shown in Fig. 7.1.

We can create a simple linear function, change the slope, and follow it with the predictor. We can alter the smoothing parameters to see how the prediction changes and adapts to the systems that it is following. The front panel of the program could look like the one shown in Fig. 7.2.

The block diagram is shown in Fig. 7.3. Shift registers are used to accumulate the past and present measurements. These measurements are stored in an array and inverted so the newest measurement is at the end of the array.

7.8 Gray Modeling and Prediction

Gray theory is a novel scientific theory originally proposed by J. Deng [12, 13] in 1982. If a system is observed from external references, it is called a black box. If the parameters and properties are well known, it is called a white system. Thus, a system

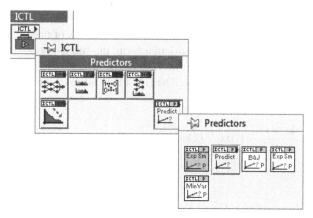

Fig. 7.1 Predictors palette at ICTL

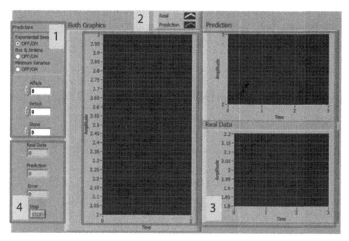

1. Controls for selecting the predictive method and their different parameters
2. Graphical predicted and real data
3. Separated graphical outputs for the prediction and real data
4. Numerical indicators for the real data and the predicted

Fig. 7.2 Front panel for the predictors example

with partially known data is called a "gray" system. The name gray is defined for these kinds of systems.

Gray theory treats any variation as gray data in a certain range and random processes are considered as gray time-varying in a certain range. It also generates data to obtain more regular generating sequences from original random data. The gray prediction employs past and presently known or indeterminate data to establish a gray model. The model can be used to predict future variations in the tendency of the output.

A specific feature of gray theory is its use of discrete-time sequences of data to build up a first-order differential equation. On a particular form, the single-variable first-order differential equation is used to model the GM (1, 1), which only uses a small portion of the data for the modeling process. The GM (1, 1) model is defined by the following equation:

$$\frac{\mathrm{d}x^{(1)}(k)}{\mathrm{d}k} + ax^{(1)}(k) = b.$$ (7.39)

7.8.1 Modeling Procedure of the Gray System

The original data is preprocessed using the accumulated generating operation (AGO) in order to decrease the random behavior of the system and to obtain the modeling information. Then, the generated data is taken to construct the model.

Algorithm 7.1

1. Let the original data be $x^{(0)}$: $x^{(0)} = \left(x^{(0)}(1), x^{(0)}(2), \ldots, x^{(0)}(n)\right)$ $n = 4, 5, \ldots$

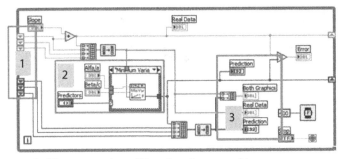

1. Three samples of the signal are used to predict
2. Selector for the desired method
3. Display of Results

Fig. 7.3 Block diagram of predictors example

Since the GM prediction is a local curve fitting extrapolation scheme, at least four data samples are required to obtain an approximate prediction. Five samples can yield better results. In addition, the prediction accuracy is not proportional to the number of samples. Additionally a forgetting term can be applied so the most recent data has more weight than the older one. A linearly increasing weighting may be applied, but an exponential form is more popular. In that case the original data series would be transformed as in (7.40), where α is the forgetting factor:

$$\alpha x^{(0)} = \left(\alpha x^{(0)}(1), \alpha^n x^{(0)}(2), \ldots, \alpha^n x^{(0)}(n)\right) \quad 0 < \alpha < 1. \quad (7.40)$$

2. Let $x^{(1)}$ be the one time AGO (1-AGO) of $x^{(0)}$: $x^{(1)} = \left(x^{(1)}(1), x^{(1)}(2), \ldots, x^{(1)}(n)\right)$, where $x^{(1)}(k) = \sum_{m=1}^{k} x^{(0)}(m) \quad k = 1, 2, \ldots m.$
3. Using least square means the model parameters $\hat{a}$ are calculated as:

$$\hat{a} = \begin{bmatrix} a \\ b \end{bmatrix} = \left(B^T B\right)^{-1} B^T y_n, \quad (7.41)$$

where

$$B = \begin{bmatrix} -1/2\left(x^{(1)}(1) + x^{(1)}(2)\right) & 1 \\ -1/2\left(x^{(1)}(2) + x^{(1)}(3)\right) & 1 \\ \vdots & \vdots \\ -1/2\left(x^{(1)}(n-1) + x^{(1)}(n)\right) & 1 \end{bmatrix} \quad (7.42)$$

$$y_n = \begin{bmatrix} x^{(0)}(2) \\ x^{(0)}(3) \\ \vdots \\ x^{(0)}(n) \end{bmatrix}. \quad (7.43)$$

4. Then the predictive function can be obtained with: $\hat{x}^{(1)}(k) = \left(x^{(0)}(1) - \frac{b}{a}\right) \cdot e^{-ak} + \frac{b}{a}.$
Then the inverse accumulated generating operation (IAGO) is used to obtain the predictive series $\hat{x}^{(0)}$: $\hat{x}^{(0)} = \left(\hat{x}^{(0)}(1), \hat{x}^{(0)}(2), \ldots, \hat{x}^{(0)}(n)\right),$
where $\hat{x}^{(0)}(k) = \hat{x}^{(1)}(k) - \hat{x}^{(1)}(k-1) \quad k = 2, 3, \ldots n$ and $\hat{x}^{(0)}(1) = \hat{x}^{(1)}(1).$

7.9 Example of a Gray Predictor Using the ICTL

The development of an example using a gray predictor is shown in this section. We first enter the number of samples that are going to be used to create the model and the points of the signal to be predicted. The front panel is shown in Fig. 7.4.

A 1D interpolator called **Automatic_1D-Array_Interpolator.vi** is used to create information between the introduced points of the signal (Fig. 7.5). We need to

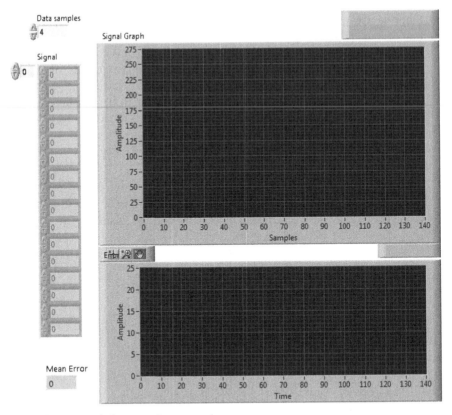

Fig. 7.4 Front panel of gray predictor example

Fig. 7.5 Diagram of
the **Automatic_1D-
Array_Interpolator.vi**

Fig. 7.6 Diagram of the
K-Step Gray Prediction.vi

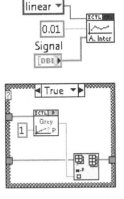

accumulate the desired samples of the signal in order to update the parameters of
the model. Next, we will introduce them to the **K-Step Gray Prediction.vi** that
executes the prediction, as shown in Fig. 7.6.

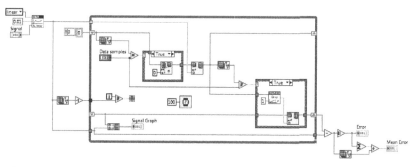

Fig. 7.7 Block diagram of gray example

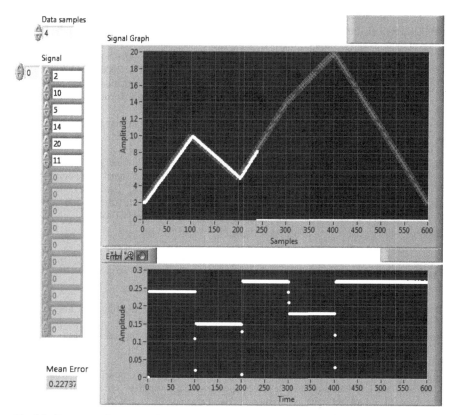

Fig. 7.8 Gray example program in action

The complete block diagram of the code is shown in Fig. 7.7. The program running would look like the one in Fig. 7.8. We will be able to see that the predictor starts taking samples of the signal (the gray one in the background) to be predicted (white) and reconstructs it.

References

1. Shen S (2008) A novel prediction method for stock index applying gray theory and neural networks. Proceedings of the 7th International Symposium on Operations Research and its Applications (ISORA 2008), China, Oct 31–Nov 3 2008, pp 104–111
2. Goa BH, Teo HP (2000) Forecasting construction industry demand, price and productivity in Singapore: the Box–Jenkins approach. Constr Manage Econ 18(5):607–618
3. Paz M, Quintero E, Fernández R (2004) Generalized minimum variance with pole assignment controller modified for practical applications. Proceedings of IEEE International Conference on Control Applications, Taiwan, pp 1347–1352
4. Snyder R, Koehler A, Ord J (2002) Forecasting for inventory control with exponential smoothing. Int J Forecast 18:5–18
5. Blankenhorn P, Gattani N, Del Castillo E, Ray C (2005) Time series and analysis and control of a dry kiln. Wood Fiber Sci 37(3):472–483
6. Ponce P, et al. (2007) Neuro-fuzzy controller using LabVIEW. Paper presented at the Intelligent Systems and Control Conference by IASTED, Cambridge, MA, 19–21 Nov 2007
7. Ponce P, Saint Martín R (2006) Neural networks based on Fourier series. Proceedings of IEEE International Conference on Industrial Technology, India, 15–17 Dec 2006, pp 2302–2307
8. Ramirez-Figueroa FD, Mendez-Cisneros D (2007) Neuro-fuzzy navigation system for mobile robots. Dissertation, Electronics and Communications Engineering, Tecnológico de Monterrey, México, May 22, 2007
9. Ponce P, et al. (2006) A novel neuro-fuzzy controller based on both trigonometric series and fuzzy clusters. Proceedings of IEEE International Conference an Industrial Technology, India, 15–17 Dec, 2006
10. Kanjilal PP (1995) Adaptive prediction and predictive control. IEE, London
11. Wellstead PE, Zarrop MB (1991) Self-tuning systems control and signal processing. Wiley, New York
12. Deng JL (1982) Control problems of gray systems. Syst Control Lett 1(5):288–294
13. Deng JL (1989) Introduction to gray systems theory. J Gray Syst 1(1):1–24

Futher Reading

Bohlin T, Graebe SF (1995) Issues in nonlinear stochastic grey-box identification. International Journal of Adaptive Control and Signal Processing 9:465–2013490

Bowerman B, O'Connel R (1999) Forecasting and time series: an applied approach, 3rd edn. Duxbury Press, Pacific Grove, CA

Fullér R (2000) Introduction to neuro-fuzzy systems. Physia, Heidelberg

Holst J, Holst U, Madsen H, Melgaard H (1992) Validation of grey box models. In L. Dugard, M. M'Saad, and I. D. Landau (Eds.), Selected papers from the fourth IFAC symposium on adaptive systems in control and signal processing, Oxford, Pergamon Press, 407–2013414

Hsu Y-T, Yeh J (1999) Grey-neural forecasting system. Proceedings of 5th International Symposium on Signal Processing and its Applications (ISSPA 1999), Aug 1999

Huang S-J, Huang C-L (2000) Control of an inverted pendulum using grey prediction model. IEEE Trans Ind Appl 36(2):452–458

Jang J-SR, Sun C-T, Mitzutani E (1997) Introduction to neuro-fuzzy and soft computing. Neuro-fuzzy and soft computing. Prentice Hall, New York

Siegwart R, Nourbakhsh IR (2004) Introduction to autonomous mobile robots. MIT Press, Cambridge, MA

Index